PRAISE FOR MICHAEL BARTON

From "*The Power of Regenerative Agriculture:*"

This is your go-to book to help inspire and educate yourself about the possibility of how you can play a role in the interconnection of farming systems with the entirety of the ecological system.

— D. G. FARNSWORTH

Well done book. I liked the chapter summaries at the end of each chapter. Chapter 6, was spot on in describing the challenges and barriers to implementing regenerative agriculture. The author has a passion for this subject and has provided very clear and concise data. The Afterword ends the book on a positive note and a view of hope that we'll get there.

— SRP

These guides to being environmentally responsible often come across as a little preachy. This is not one of those. It's informative and well-written, with a cheerful attitude toward ecologically friendly farming methods.

— LWHIT

RAINWATER HARVESTING FOR YOUR HOME

RAINWATER HARVESTING FOR YOUR HOME

A COMPREHENSIVE GUIDE TO SUSTAINABLE RAINWATER COLLECTION FOR HOMEOWNERS

SUSTAINABLE LIVING AT HOME

MICHAEL BARTON

Book
Bound Studios

The information contained in this book is based on the author's personal experiences and research. While every effort has been made to ensure the accuracy of the information presented, the author and publisher cannot be held responsible for any errors or omissions.

This book is intended for general informational purposes only and is not a substitute for professional medical, legal, or financial advice. If you have specific questions about any medical, legal, or financial matters, you should consult with a qualified healthcare professional, attorney, or financial advisor.

Book Bound Studios is not affiliated with any product or vendor mentioned in this book. The views expressed in this book are those of the author and do not necessarily reflect the views of Book Bound Studios.

To all those who strive to live sustainably and cherish our planet, this book is dedicated to you. May it serve as a guide and inspiration for harnessing the power of rainwater and reducing our reliance on precious natural resources. Let us work together to build a brighter, greener future for generations to come.

Water is the driving force of all nature.

— LEONARDO DA VINCI

CONTENTS

INTRODUCTION

Welcome to *"Rainwater Harvesting for Your Home,"* where we'll explore the fascinating world of harvesting rainwater! As we become more aware of our environmental impact, we must find sustainable living ways. One way we can do this is by **harvesting rainwater.**

In this chapter, we'll dive into the importance of rainwater harvesting and how it can contribute to a more sustainable lifestyle. We'll cover the essential components of a rainwater harvesting system, so you can understand how it all works. Then, you'll learn how to assess your home's potential for rainwater harvesting, so you can make informed decisions about whether or not it's right for you.

We'll also explore legal and regulatory considerations for rainwater harvesting, so you can be sure you're following the rules and regulations in your area. Finally, we'll discuss the many benefits of rainwater harvesting and how it can help us move toward a greener future.

So, whether you're a seasoned environmentalist or just starting your journey toward sustainable living, this book has something for you. Let's get started!

The Importance of Rainwater Harvesting for Sustainable Living

In today's rapidly changing world, sustainable living has become more critical. As we strive to reduce our ecological footprint and conserve our precious natural resources, rainwater harvesting emerges as a practical and effective solution. This section will delve into the significance of rainwater harvesting and its role in promoting a sustainable lifestyle.

Rainwater harvesting, at its core, is collecting, storing, and utilizing rainwater for various purposes, such as irrigation, watering plants, flushing toilets, and even drinking when adequately treated. By capturing and reusing this valuable resource, we can significantly reduce reliance on traditional water sources, such as municipal water supplies, wells, and rivers. This, in turn, has numerous environmental, economic, and social benefits.

First and foremost, rainwater harvesting helps conserve water, one of our most vital natural resources. With climate change causing unpredictable weather patterns and increasing instances of drought, water scarcity has become a pressing global issue. By harvesting rainwater, we can alleviate the pressure on our existing water sources and ensure a more sustainable supply for future generations.

Moreover, rainwater harvesting reduces the demand for our overburdened stormwater management systems. By capturing rainwater before it enters the storm drains, we can minimize the risk of urban flooding and decrease the pollutants that flow into our rivers, lakes, and oceans. This protects our aquatic ecosystems and improves the overall water quality for humans and wildlife.

From an economic standpoint, rainwater harvesting can lead to significant cost savings for homeowners. By supplementing or replacing the need for municipal water, households can reduce their monthly water bills and become more self-sufficient. Additionally, implementing rainwater harvesting systems can increase property values, as potential buyers see them as an attractive and eco-friendly feature.

Rainwater harvesting also plays a crucial role in promoting social equity. In many parts of the world, clean water is a luxury that millions cannot afford. By implementing rainwater harvesting systems in these communities, we can help bridge the gap between the water-rich and water-poor, ensuring that everyone has the opportunity to live a healthy and dignified life.

Lastly, rainwater harvesting encourages a greater connection to and appreciation for the natural world. By actively engaging in the process of collecting and using rainwater, we become more aware of our environment and the impact our actions have on it. This heightened sense of responsibility can inspire us to adopt other sustainable practices, ultimately contributing to a greener and more resilient future for our planet.

In conclusion, rainwater harvesting is an essential component of sustainable living. By embracing this practice, we can conserve water, protect our ecosystems, save money, promote social equity, and foster a deeper connection with the environment. As we move forward in our quest for a more sustainable world, rainwater harvesting will undoubtedly play a pivotal role in shaping our collective future.

Understanding the Basic Components of a Rainwater Harvesting System

Rainwater harvesting is a simple yet effective method of collecting and storing rainwater for later use. It is an environmentally friendly practice that promotes sustainable living by reducing our reliance on municipal water supplies. This section will explore the fundamental components of a rainwater harvesting system and how they work together to provide a reliable water source for your home.

A typical rainwater harvesting system consists of the following key components:

Catchment Surface: The catchment surface is where rainwater is collected, usually on your home's roof. The size and material of your roof will determine the amount of water you can collect. Generally, larger roofs made of smooth, non-porous materials like metal or tiles are more effective at capturing rainwater.

Gutters and Downspouts: These are essential for directing the rainwater from your roof to the storage tank. Gutters should be properly maintained and cleaned regularly to prevent debris from clogging the system and reducing efficiency. Downspouts should be designed to minimize splashing and overflow, ensuring the maximum amount of water is collected.

Leaf Screens and First Flush Diverters: Leaf screens are installed at the entrance of the downspouts to prevent leaves and other debris from entering the system. First, flush diverters separate the initial dirty water from the rest of the rainwater, ensuring that only clean water enters the storage tank. This helps maintain the stored water's quality and reduces the need for treatment.

Storage Tank: The storage tank is the heart of the rainwater harvesting system. It is where the collected rainwater is stored for later use. Storage tanks can be made from various materials, such as plastic, concrete, or metal, and come in different sizes to suit your needs. The tank size will depend on the water you wish to store and the space available on your property.

Filtration and Treatment: Depending on the intended use of the harvested rainwater, you may need to install a filtration and treatment system—this range from simple sediment filters to more advanced UV or chemical treatment systems. Filtration and treatment are essential if you use the harvested water for drinking or other domestic purposes.

Delivery System: The delivery system distributes the stored rainwater to various points of use within your home. Depending on your needs and preferences, this can include a simple gravity-fed system or a more complex pump-driven system. The delivery system should be designed to ensure that the water pressure and flow rates are suitable for your intended uses.

Overflow and Drainage: It is essential to have a proper overflow and drainage system in place to prevent your storage tank from overflowing during heavy rainfall. This can be achieved by installing an overflow pipe that directs excess water away from the tank and into a safe drainage area.

By understanding the essential components of a rainwater harvesting system, you can make informed decisions when designing and installing your system. The next section will discuss assessing your home's rainwater harvesting potential and determining the most suitable method for your needs.

Assessing Your Home's Rainwater Harvesting Potential

Rainwater harvesting is an excellent way to reduce water consumption, save money on water bills, and contribute to a more sustainable lifestyle. However, before you dive into the world of rainwater harvesting, it's essential to assess your home's potential for collecting and utilizing rainwater effectively. In this section, we will guide you through evaluating your home's rainwater harvesting potential, considering factors such as roof size, rainfall patterns, and water usage.

Roof Size and Material

The first step in assessing your home's rainwater harvesting potential is to determine the size and material of your roof. The roof area is crucial in determining the amount of rainwater you can collect, as a larger surface area will collect more water. To calculate the size of your roof, measure the length and width of your home and multiply these figures together.

The material of your roof also plays a significant role in the efficiency of your rainwater harvesting system. Ideally, your roof should be made of a non-toxic material that does not leach harmful chemicals into the collected water. Some suitable materials include metal, slate, and clay tiles. If your roof is made of less desirable material, such as asphalt shingles, consider installing a pre-filtration system to ensure the collected water is safe.

Rainfall Patterns

The next factor to consider is the average annual rainfall in your area. This information can usually be found through local weather stations or online resources. Knowing the average rainfall will help you estimate the amount of water you can collect and determine the storage tank size needed to accommodate your needs.

Remember that rainfall patterns can vary significantly from year to year, so planning for both average and extreme conditions is essential. In addition, consider the seasonal distribution of rainfall, as this can affect the efficiency of your system and the need for additional water storage during drier months.

Water Usage

To maximize the benefits of your rainwater harvesting system, it's essential to have a clear understanding of your household's water usage. Begin by analyzing your water bills to determine your average monthly consumption. Then, consider how to use rainwater in your daily activities, such as watering plants, flushing toilets, and even washing clothes. By identifying the areas where rainwater can replace or supplement your current water usage, you can better estimate the storage tank size and filtration system needed to meet your needs.

Site Assessment

Finally, conduct a thorough site assessment of your property to identify potential challenges or limitations for installing a rainwater harvesting system. Some factors to consider include the slope of your land, the location of your home, your property boundaries, and the presence of any underground utilities or obstacles that may interfere with the installation of your system.

By assessing your home's rainwater harvesting potential, you can ensure that your investment in a rainwater harvesting system is well-planned and tailored to your specific needs. In the next section, we will discuss the legal and regulatory considerations for rainwater harvesting, helping you navigate the process of obtaining the necessary permits and approvals for your system.

Legal and Regulatory Considerations for Rainwater Harvesting

As we delve deeper into the world of rainwater harvesting, it's essential to address the legal and regulatory aspects that govern this practice. While rainwater harvesting is a sustainable and eco-friendly solution to water scarcity, it is crucial to ensure that your system adheres to the laws and regulations in your area. In this section, we will discuss the legal landscape surrounding rainwater harvesting and provide guidance on navigating the regulatory framework to ensure your system is compliant.

Understanding the Legal Landscape

The legal landscape for rainwater harvesting varies significantly from one jurisdiction to another. In some regions, rainwater harvesting is encouraged through incentives and rebates. In others, it may be subject to strict regulations or even prohibited. This diversity in regulations stems from differences in water rights, local climate conditions, and the perceived impact of rainwater harvesting on the environment and existing water infrastructure.

To ensure that your rainwater harvesting system is legally compliant, familiarize yourself with your area's laws and regulations. This may involve researching local building codes, zoning ordinances, and water rights laws and consulting with local authorities and experts in the field.

Navigating the Permitting Process

In many jurisdictions, installing a rainwater harvesting system requires obtaining permits from local authorities. The permitting process typically involves submitting detailed plans of your proposed system, including information on the system's components, capacity, and intended use. Additionally, you may be required to demonstrate that your system meets specific safety, health, and environmental standards.

To navigate the permitting process smoothly, consulting with professionals experienced in rainwater harvesting system design and installation is advisable. These experts can help you develop a compliant system and guide you through the necessary steps to obtain the required permits.

Compliance with Building Codes and Zoning Ordinances

Your rainwater harvesting system must comply with local building codes and zoning ordinances. These regulations often dictate the allowable size, location, and appearance of rainwater harvesting systems and the materials and construction methods used.

For example, some building codes may require that rainwater storage tanks be constructed of specific materials or meet certain structural standards. Similarly, zoning ordinances may restrict the placement of rainwater harvesting systems in certain areas or require that they be screened from view.

To ensure compliance with these regulations, consult with local building and zoning officials and seek professional assistance in designing and installing your system.

Water Rights Considerations

In some jurisdictions, water rights laws may impact your ability to harvest rainwater. These laws govern the ownership and use of water resources and, in some cases, may restrict the collection and use of rainwater.

Suppose you live in an area with strict water rights laws. In that case, it is crucial to research these regulations and consult with legal experts to determine if rainwater harvesting is permissible and under what conditions.

Staying Informed and Adapting to Changes

As the importance of sustainable water management continues to gain recognition, the legal and regulatory landscape for rainwater harvesting is likely to evolve. To ensure ongoing compliance, stay informed about changes in laws and regulations in your area and be prepared to adapt your system as needed.

In conclusion, understanding and navigating the legal and regulatory considerations for rainwater harvesting is crucial to implementing a successful and compliant system. By researching local laws, consulting with experts, and staying informed about regulation changes, you can confidently embrace rainwater harvesting as a sustainable solution for your home and contribute to a greener future.

Embracing Rainwater Harvesting for a Greener Future

As we reach the end of this introduction, it is essential to take a moment to reflect on the significance of this sustainable practice in our daily lives. By now, you have been introduced to the importance of rainwater harvesting for sustainable living, the essential components of a rainwater harvesting system, how to assess your home's potential for rainwater harvesting, and the legal and regulatory considerations involved in implementing such a system. With this knowledge, you are ready to continue to the next chapter.

Embracing rainwater harvesting is not just about conserving water and reducing your environmental footprint. It is also about adopting a mindset that values our planet's resources and seeks to preserve them for future generations. By implementing a rainwater harvesting system in your home, you are taking a significant step towards sustainable living and contributing to the global effort to combat water scarcity and climate change.

Moreover, rainwater harvesting offers numerous benefits beyond environmental sustainability. Utilizing this natural resource can reduce your reliance on municipal water supplies, lower your water bills, and even improve the quality of the water you use for various purposes around your home. Additionally, rainwater harvesting systems can be customized to suit your specific needs and preferences, allowing you to create a solution that works best for your household.

As you embark on your rainwater harvesting journey, remember that every drop counts. Even small-scale systems can significantly impact conserving water and promoting a more sustainable lifestyle. Furthermore, sharing your experiences and knowledge with others can inspire your community to adopt similar practices and create a ripple effect of positive change.

In conclusion, rainwater harvesting is a practical, efficient, and environmentally responsible solution to address growing water scarcity and climate change concerns. By embracing and incorporating this practice into your home, you can play a crucial role in safeguarding our planet's precious resources and fostering a greener future for all. So, continue reading and take the plunge – the rewards of rainwater harvesting await you!

1

THE IMPORTANCE OF RAINWATER HARVESTING FOR YOUR HOME

A serene forest glen in the midst of a gentle rain, with water droplets glistening on the leaves.

In today's rapidly changing world, where the demand for natural resources is ever-increasing, adopting sustainable practices that can help preserve our environment and ensure a healthy future for future generations is crucial. One such practice that has gained significant attention in recent years is **rainwater harvesting.** As the name suggests, rainwater harvesting is collecting, storing, and utilizing rainwater for various purposes, such as irrigation, gardening, and even domestic use after proper treatment. This section will explore the importance of rainwater harvesting for your home and how it can contribute to a more sustainable and eco-friendly lifestyle.

Rainwater harvesting is not new; it has been practiced for centuries across different civilizations. However, with the advent of modern plumbing systems and the easy availability of water, this ancient practice was largely forgotten. But as the world faces increasing water scarcity and environmental challenges, rainwater harvesting has once again emerged as a viable and essential solution.

The importance of rainwater harvesting for your home cannot be overstated. From environmental benefits to financial advantages, rainwater harvesting offers a multitude of reasons for homeowners to consider implementing this practice. In the following sections, we will delve deeper into these benefits, discussing how rainwater harvesting can improve water security, enhance the health of your landscape and garden, and contribute to a more sustainable future for all.

By the end of this chapter, you will have a comprehensive understanding of the significance of rainwater harvesting for your home. You will be inspired to take action toward a greener and more sustainable lifestyle. So, let's embark on this journey together and discover the incredible potential of rainwater harvesting in transforming our lives and the world around us.

Environmental Benefits of Rainwater Harvesting

Rainwater harvesting is a practical solution for homeowners looking to reduce their water bills and increase their self-sufficiency. It also offers a range of environmental benefits that contribute to a healthier planet. In this section, we will explore how rainwater harvesting can positively impact the environment and help preserve our precious water resources.

First and foremost, rainwater harvesting **reduces the demand for our existing water supply systems.** By collecting and using rainwater, homeowners can significantly decrease their reliance on municipal water sources, which are often overburdened and subject to shortages during periods of drought. This reduced demand helps conserve water resources and ensures enough water is available for essential uses such as drinking, sanitation, and agriculture.

Another environmental benefit of rainwater harvesting is the **reduction of stormwater runoff.** When rain falls on impervious surfaces such as roofs, driveways, and streets, it can quickly accumulate and cause flooding, erosion, and water pollution. By capturing rainwater before it becomes runoff, homeowners can help mitigate these issues and protect local waterways from contamination. This is particularly important in urban areas, where stormwater runoff can carry pollutants such as oil, pesticides, and heavy metals into rivers and streams, harming aquatic life and degrading water quality.

Rainwater harvesting also contributes to **energy conservation.** Treating and distributing water requires a significant amount of energy. By using rainwater instead of relying solely on treated water, homeowners can help reduce the energy consumption associated with water provision. This, in turn, helps to decrease greenhouse gas emissions and our overall carbon footprint.

Furthermore, rainwater is a naturally soft water source, containing fewer minerals and chemicals than treated water. This can benefit your garden and landscape, **reducing the risk of chemical buildup** in the soil and promoting healthier plant growth. Additionally, using rainwater for irrigation can help prevent the overuse of chemical fertilizers and pesticides, which can harm the environment and pollute our waterways.

In conclusion, the environmental benefits of rainwater harvesting are numerous and far-reaching. By embracing this sustainable practice, homeowners can play an active role in conserving water resources, protecting local ecosystems, and promoting a healthier environment for future generations. Furthermore, as we continue to face the challenges of climate change and water scarcity, rainwater harvesting offers a practical and eco-friendly solution that can significantly impact our collective efforts to preserve the planet.

Financial Advantages of Collecting Rainwater

In today's world, where the cost of living is constantly rising, finding ways to save money and reduce expenses is essential. One often overlooked method of cutting costs is rainwater harvesting. By collecting and utilizing rainwater, homeowners can enjoy significant financial advantages. In this section, we will explore how rainwater harvesting can lead to monetary savings and increased financial stability.

First and foremost, rainwater harvesting **reduces your reliance on municipal water supplies**. By capturing and storing rainwater, you can decrease your consumption of treated water from the city or town. This reduction in usage directly translates to lower water bills, leaving more money in your pocket. Additionally, as water rates continue to rise, the savings from rainwater harvesting will only become more substantial.

Another financial advantage of rainwater harvesting is the potential for reduced stormwater management costs. In many urban areas, stormwater fees are levied on property owners based on the impervious surface of their property. By implementing rainwater harvesting systems, you can reduce the volume of stormwater runoff from your property, lowering your stormwater fees.

Moreover, rainwater harvesting can lead to savings in energy costs. Municipal water treatment and distribution require a significant amount of energy, which is often reflected in the cost of water. By using rainwater for non-potable purposes, such as flushing toilets and irrigation, you can reduce the demand for treated water and, in turn, the energy required to produce it. This can lead to lower energy bills and a smaller carbon footprint.

Furthermore, rainwater harvesting can contribute to increased property value. A well-designed and maintained rainwater harvesting system can attract potential homebuyers, particularly those who are environmentally conscious or interested in self-sufficiency. By investing in a rainwater harvesting system, you can increase the resale value of your home, making it a wise financial decision in the long run.

Finally, government incentives and rebates are often available for homeowners who invest in rainwater harvesting systems. These incentives can offset the initial costs of installing a system, making it a more affordable and financially viable option: local research programs and rebates to take full advantage of these opportunities.

In conclusion, the financial advantages of collecting rainwater are numerous and wide-ranging. From reduced water and energy bills to increased property value and government incentives, rainwater harvesting is a wise investment for homeowners looking to save money and contribute to a more sustainable future.

Improving Water Security and Self-Sufficiency

Water security and self-sufficiency are becoming increasingly important for homeowners in today's rapidly changing world. As climate change continues to disrupt traditional water sources and urbanization puts pressure on existing infrastructure, it is crucial to explore alternative methods to ensure a consistent and reliable water supply for our homes. Rainwater harvesting is one such method that can significantly contribute to improving water security and self-sufficiency.

Water security refers to the availability of an adequate quantity and quality of water for various household needs, such as drinking, cooking, and sanitation. By collecting and storing rainwater, homeowners can reduce their dependence on municipal water supplies, which may be subject to shortages or contamination. This is particularly important in regions where water scarcity is a pressing issue, as it helps to alleviate the burden on already strained resources.

Moreover, rainwater harvesting systems can serve as a **backup water source during emergencies or disruptions** in the public water supply. In a natural disaster or infrastructure failure, having a reserve of stored rainwater can provide a vital lifeline for families, ensuring they have access to clean water for their essential needs.

Self-sufficiency, on the other hand, refers to the ability of a household to meet its water requirements independently without relying on external sources. By harvesting rainwater, homeowners can significantly reduce their water bills and gain greater control over their water usage. This leads to financial savings and fosters a sense of empowerment and self-reliance as individuals take responsibility for their water management.

Additionally, rainwater harvesting systems can be tailored to suit each household's specific needs and preferences. Homeowners can choose from various collection methods, storage capacities, and treatment options, allowing them to design a system that best meets their unique requirements. This flexibility further enhances the self-sufficiency aspect of rainwater harvesting, enabling individuals to adapt their systems to changing circumstances or evolving needs.

In conclusion, rainwater harvesting offers a practical and sustainable solution for improving water security and self-sufficiency in our homes. By embracing this method, homeowners can safeguard their water supply and contribute to a more resilient and resource-efficient future.

Enhancing Landscape and Garden Health

One of the most noticeable and rewarding advantages of rainwater harvesting for your home is its positive impact on the health and vitality of your landscape and garden. By utilizing the natural resource that falls from the sky, you can provide your plants with the nourishment they need to thrive while promoting a sustainable and eco-friendly approach to gardening.

Rainwater is an excellent source of hydration for your plants compared to tap water. This is because rainwater is free of chemicals, such as chlorine and fluoride, commonly found in municipal water supplies. These chemicals can harm the delicate balance of your garden's ecosystem, potentially causing damage to your plants and soil. In contrast, rainwater is naturally soft and contains essential nutrients and minerals beneficial for plant growth.

When you harvest rainwater for your garden, you are not only providing your plants with a healthier water source but also helping to reduce the demand for our overburdened water supply systems. This is particularly important in areas where water scarcity is a growing concern. In addition, by using rainwater to irrigate your landscape, you can contribute to conserving this precious resource and reducing the strain on our environment.

Another advantage of rainwater harvesting is that it can help **improve the overall soil quality** in your garden. As rainwater percolates through the soil, it helps to break up compacted earth, allowing for better aeration and drainage. In addition, this process encourages the growth of beneficial microorganisms and earthworms, which are essential for maintaining a healthy soil structure. In turn, this leads to stronger root systems and more robust plants.

In addition to the benefits for your plants and soil, rainwater harvesting can contribute to a more attractive and diverse landscape. By designing your rainwater collection system to include features such as rain gardens, swales, or permeable paving, you can create an aesthetically pleasing environment supporting various plant species and wildlife. This enhances the beauty of your outdoor space and promotes biodiversity and ecological balance.

In conclusion, incorporating rainwater harvesting into your home's landscape and garden design can bring numerous benefits to the health and vitality of your plants, soil, and overall ecosystem. By embracing this sustainable practice, you can contribute to conserving our planet's resources while enjoying a thriving and beautiful outdoor sanctuary.

Embracing Rainwater Harvesting for a Sustainable Future

In conclusion, rainwater harvesting is an essential practice that benefits homeowners and the environment. By embracing this sustainable approach to water management, we can contribute to a greener and more resilient future for our communities and the planet.

Throughout this chapter, we have explored the environmental benefits of rainwater harvesting, such as reducing the strain on our water resources, minimizing stormwater runoff, and promoting groundwater recharge. These advantages help preserve our ecosystems and contribute to a healthier environment for future generations.

Financially, rainwater harvesting systems can significantly save water bills, making it an attractive investment for homeowners. As water scarcity and the cost of water continue to rise, these savings will become increasingly valuable. Moreover, government incentives and rebates can further sweeten the deal, making rainwater harvesting an even more appealing option.

Water security and self-sufficiency are also vital aspects of rainwater harvesting. By collecting and storing rainwater, homeowners can reduce their dependence on municipal water supplies and ensure a reliable water source during drought or water restrictions. In addition, this increased self-reliance can provide peace of mind and a sense of empowerment, knowing that you are taking control of your water needs.

Furthermore, rainwater harvesting can enhance the health and beauty of your landscape and garden. The natural, chemical-free quality of rainwater is ideal for nurturing plants, leading to more vibrant and flourishing green spaces around your home. This not only improves the aesthetics of your property but also contributes to the overall well-being of the environment.

In summary, rainwater harvesting is a practical and sustainable solution that can benefit your home and the environment. By adopting this eco-friendly practice, you can actively conserve our precious water resources, save money, and promote a greener future for all. So, let us embrace rainwater harvesting and positively impact our world, one drop at a time.

Chapter Summary

1. Rainwater harvesting is a sustainable practice that involves collecting, storing, and utilizing rainwater for various purposes, such as irrigation, gardening, and domestic use after proper treatment.

2. Environmental benefits of rainwater harvesting include reducing the demand on existing water supply systems, minimizing stormwater runoff, conserving energy, and providing a naturally soft water source for gardens and landscapes.
3. Financial advantages of collecting rainwater include lower water bills, reduced stormwater management costs, potential energy savings, increased property value, and government incentives and rebates.
4. Rainwater harvesting can improve water security and self-sufficiency by reducing dependence on municipal water supplies and serving as a backup water source during emergencies or disruptions.
5. Utilizing rainwater for landscape and garden irrigation can enhance plant health and soil quality. It is free of chemicals found in treated water and contains essential nutrients and minerals.
6. Rainwater harvesting systems can be tailored to suit each household's specific needs and preferences, allowing for flexibility and adaptability.
7. Incorporating rainwater harvesting into landscape design can create a more attractive and diverse environment, supporting various plant species and wildlife.
8. Embracing rainwater harvesting contributes to a greener and more resilient future for communities and the planet, conserving water resources, promoting biodiversity, and fostering a more sustainable lifestyle.

2

UNDERSTANDING THE RAINWATER HARVESTING PROCESS

A rainwater tank made entirely out of recycled materials, sitting in the middle of a lush garden.

Rainwater harvesting is an ancient practice utilized by various civilizations throughout history. Today, it continues to gain popularity as a sustainable and eco-friendly solution to water scarcity and the increasing demand for clean water. This chapter will delve into the fascinating world of rainwater harvesting, exploring its importance, components, and the steps required to implement a successful system for your home.

At its core, rainwater harvesting is collecting, storing, and utilizing rainwater for various purposes, such as irrigation, flushing toilets, and even potable water when adequately treated. This simple yet effective method allows homeowners to reduce their reliance on municipal water supplies and decrease their water bills while contributing to a greener and more sustainable environment.

The beauty of rainwater harvesting lies in its **adaptability**. Whether you live in an urban apartment or a sprawling suburban home, a rainwater harvesting system can be tailored to suit your needs. The possibilities are endless, from small-scale systems that collect water in barrels to more complex setups that integrate with your home's plumbing.

In the following sections, we will discuss the importance of rainwater harvesting for your home and the benefits it can provide. We will also examine the various components and design elements of a rainwater harvesting system, guiding you through the steps necessary to implement your system. Finally, we will cover the essential maintenance and troubleshooting aspects to ensure that your rainwater harvesting system remains efficient and effective for years.

So, let's embark on this exciting journey towards sustainable water management and discover how rainwater harvesting can transform your home and your relationship with this precious resource.

The Importance of Rainwater Harvesting for Your Home

Rainwater harvesting is not just an eco-friendly practice but a vital step towards sustainable living and responsible water management. Yet, as homeowners, we often take our water supply for granted, not realizing our consumption's impact on the environment and our wallets. This section will delve into the importance of rainwater harvesting for your home, exploring its numerous benefits and how it contributes to a greener, more sustainable future.

First and foremost, rainwater harvesting helps **conserve a precious resource: water.** With climate change and increasing urbanization, water scarcity is becoming a pressing issue worldwide. Collecting and utilizing rainwater directly reduces your reliance on municipal water supplies, often sourced from overburdened rivers, lakes, and aquifers. This, in turn, helps preserve these water sources for future generations and supports the overall health of our ecosystems.

Another significant advantage of rainwater harvesting is the reduction of your water bills. As you rely more on harvested rainwater for household needs, such as gardening, flushing toilets, and even laundry, you will notice a considerable decrease in your monthly water expenses. This financial benefit appeals to homeowners looking to cut costs and invest in long-term, sustainable solutions.

Rainwater harvesting also reduces stormwater runoff, which can cause soil erosion, flooding, and pollution in our waterways. By capturing rainwater on your property, you are minimizing the volume of water that flows into storm drains and sewer systems. This helps protect your home from potential water damage and benefits the environment by reducing the pollutants that enter our rivers, lakes, and oceans.

Furthermore, rainwater is naturally soft and chemical-free, making it an ideal choice for gardening and irrigation. Plants thrive on rainwater, as it is free of the salts, minerals, and other impurities commonly found in tap water. This results in healthier, more vibrant gardens and landscapes, which can also increase your property's aesthetic appeal and value.

Lastly, rainwater harvesting promotes a sense of environmental stewardship and self-sufficiency. By implementing a rainwater collection system, you are actively managing your water resources and reducing your ecological footprint. This benefits the planet and instills a sense of pride and accomplishment in your efforts to live a more sustainable lifestyle.

In conclusion, rainwater harvesting is essential for homeowners who wish to conserve water, save money, and contribute to a healthier environment. By understanding the importance of rainwater harvesting and incorporating it into your daily life, you are taking a significant step towards sustainable water management and a greener future.

Components and Design of a Rainwater Harvesting System

Designing an efficient rainwater harvesting system for your home is exciting and rewarding. Not only will it help you conserve water and reduce your water bills, but it will also contribute to a more sustainable environment. This section will discuss the essential components of a rainwater harvesting system and how to design one that suits your needs and preferences.

A well-designed rainwater harvesting system consists of several key components, each playing a crucial role in collecting, storing, and distributing rainwater. Let's take a closer look at these components:

Catchment Surface: The catchment surface is where rainwater is collected, such as on rooftops or other hard surfaces. The size and material of your catchment surface will directly impact the amount of water you can harvest. Therefore, you should choose a surface that is large, clean, and made of non-toxic materials.

Gutters and Downspouts: These components channel rainwater from the catchment surface to the storage tank. They should be adequately sized and installed to ensure efficient water flow and prevent clogging. Regular cleaning and maintenance of gutters and downspouts are essential to keep your system functioning optimally.

First Flush Diverter: This device helps to divert the initial flow of rainwater, which may contain contaminants and debris from the catchment surface. Doing so ensures that only clean water enters the storage tank, reducing the need for extended filtration and maintenance.

Storage Tank: The storage tank is where the collected rainwater is stored for later use. The size of your tank will depend on your water needs, available space, and budget. Common materials for storage tanks include plastic, concrete, and metal. Choosing a tank that is durable, UV-resistant, and well-insulated to prevent algae growth and maintain water quality is essential.

Filtration System: Before using the harvested rainwater, it's crucial to filter it to remove any remaining impurities. Various filtration methods are available, including sediment filters, activated carbon filters, and UV sterilization. The filtration system choice will depend on the intended use of the harvested water and the quality of the water collected.

Distribution System: Once the rainwater is filtered, it must be distributed throughout your home for various uses, such as irrigation, toilet flushing, or drinking. This can be achieved through a network of pipes, pumps, and valves that connect the storage tank to the points of use.

When designing your rainwater harvesting system, it's essential to consider factors such as local climate, rainfall patterns, and water usage habits. This will help you determine your system's appropriate size and capacity, ensuring that it meets your water needs while remaining cost-effective and environmentally friendly.

In conclusion, a well-designed rainwater harvesting system is an invaluable addition to any home. By understanding the various components and design considerations, you can create a system that saves you money and contributes to a more sustainable future.

Steps to Implementing a Rainwater Harvesting System

Now that you have a solid understanding of the rainwater harvesting process and its components, it's time to dive into the steps involved in implementing a rainwater harvesting system for your home. By following these steps, you'll be well on your way to enjoying the benefits of sustainable water management and reducing your reliance on traditional water sources.

Assess your water needs: Before you begin, it's essential to determine your household's water usage and requirements. Consider factors such as the size of your family, the types of plants in your garden, and your daily water consumption habits. This information will help you design a system that meets your specific needs.

Analyze rainfall patterns: Study the rainfall patterns in your area to determine the average annual rainfall and the distribution of rain throughout the year. This will help you estimate the amount of water you can harvest and guide you in selecting the appropriate size of your storage tank.

Choose a suitable location: Select a location for your rainwater harvesting system close to the areas where you plan to use the water, such as your garden or laundry room. Ensure the site has adequate space for the storage tank and is easily accessible for maintenance.

Design your system: With your water needs and rainfall patterns in mind, design a rainwater harvesting system that includes components such as gutters, downspouts, filters, storage tanks, and a distribution system. Consult a professional to ensure your system is designed efficiently and safely.

Obtain permits and approvals: Depending on your local regulations, you may need permits or approvals before installing your rainwater harvesting system. Check with your local authorities to ensure you comply with any necessary regulations.

Install your system: Once your design is complete and you have obtained the necessary permits, you can begin the installation process. This may involve hiring a professional installer or tackling the project yourself if you have the necessary skills and tools. Be sure to follow all safety guidelines and manufacturer recommendations during installation.

Test and adjust: After installing your system, ensure it functions correctly. This may involve checking for leaks, ensuring the filters are working properly, and verifying that the water is being distributed as intended. Then, make any necessary adjustments to optimize the performance of your system.

Educate your household: Teach your family members about the rainwater harvesting system and how to use the harvested water responsibly. Please encourage them to adopt water-saving habits to maximize the benefits of your new system.

By following these steps, you'll be well on your way to implementing a rainwater harvesting system that meets your household's needs and contributes to sustainable water management. In the next section, we'll discuss the maintenance and troubleshooting of your rainwater harvesting system to ensure its longevity and continued performance.

Maintenance and Troubleshooting of Your Rainwater Harvesting System

As with any home system, proper maintenance and troubleshooting are essential to ensure the longevity and efficiency of your rainwater harvesting system. In this section, we will delve into the necessary steps to keep your system running smoothly and address any potential issues that may arise.

Regular maintenance of your rainwater harvesting system is crucial for its optimal performance. By following a consistent maintenance schedule, you can prevent potential problems, ensure the quality of the harvested water, and prolong the life of your system components. Here are some essential maintenance tasks to keep in mind:

Inspect and clean gutters and downspouts: Regularly check your gutters and downspouts for debris, such as leaves and twigs, which can obstruct the flow of rainwater. Cleaning them at least twice a year, or more frequently if you live in an area with heavy foliage, will help maintain the efficiency of your system.

Check the roof and catchment surface: Periodically inspect your roof and catchment surface for any damage or wear that could affect the quality of the harvested rainwater. Repair any issues promptly to prevent contamination or leakage.

Monitor the pre-filtration system: Examine your pre-filtration systems for clogs or damage, such as leaf screens and first flush diverters. Clean and replace components to ensure your system effectively filters out debris and contaminants.

Inspect storage tanks and connections: Regularly check your storage tanks for leaks, cracks, or signs of wear. Additionally, examine all connections and seals to ensure they remain watertight. Repair or replace any damaged components as necessary.

Test water quality: Periodically test the quality of your harvested rainwater to ensure it meets the necessary standards for its intended use. If you notice any changes in water quality, investigate the cause and take corrective action as needed.

Maintain pumps and treatment systems: If your rainwater harvesting system includes pumps or treatment systems, follow the manufacturer's recommended maintenance schedule to ensure proper functioning.

Despite diligent maintenance, issues may still arise with your rainwater harvesting system. Troubleshooting these problems promptly can help prevent further damage and ensure the continued efficiency of your system. Here are some common issues and their potential solutions:

Reduced water flow: If you notice a decrease in the flow of harvested rainwater, check for clogs or obstructions in your gutters, downspouts, and pre-filtration system. Additionally, inspect your storage tank for sediment buildup, which can impede water flow.

Water quality issues: If your harvested rainwater has an unpleasant odor, taste, or appearance, investigate the source of the contamination. This may include damaged roofing materials, pollution from nearby sources, or inadequate filtration. Then, address the issue accordingly to restore water quality.

Leaks or overflow: Inspect the tank and connections for damage if you observe leaks or overflow from your storage tank. Additionally, ensure that your tank is adequately filled and your overflow system functions correctly.

By following a regular maintenance schedule and addressing any issues promptly, you can ensure the longevity and efficiency of your rainwater harvesting system. In doing so, you are embracing sustainable water management at home and contributing to a greener and more resilient future for our planet.

Embracing Sustainable Water Management at Home

In conclusion, rainwater harvesting is an essential and sustainable approach to water management that every homeowner should consider implementing. By embracing this eco-friendly practice, you contribute to conserving our precious water resources, reducing your dependence on municipal water supplies, lowering your water bills, and promoting a healthier environment for future generations.

Throughout this chapter, we have explored the importance of rainwater harvesting for your home, delved into the components and design of an effective rainwater harvesting system, and discussed the steps to implementing and maintaining such a system. With this knowledge, you can make informed decisions about incorporating rainwater harvesting into your home's water management strategy.

As you embark on this journey towards sustainable water management, remember that every drop of rainwater you collect and utilize is a step towards a greener and more self-sufficient home. By adopting rainwater harvesting, you benefit your household, set a positive example for your community, and inspire others to follow suit.

In a world where water scarcity is becoming increasingly pressing, we all must take responsibility for our water usage and strive to make a difference. Rainwater harvesting is a practical and effective way to do just that. By implementing this system in your home, you play a vital role in preserving our planet's most precious resource.

So, embrace sustainable water management at home – for your wallet, community, and the environment. With every raindrop you collect, you make a lasting impact and contribute to a more sustainable future.

Chapter Summary

1. Rainwater harvesting is an ancient, sustainable, and eco-friendly practice that helps conserve water, reduce reliance on municipal water supplies, and decrease water bills.
2. The importance of rainwater harvesting lies in its contribution to preserving water resources, reducing water bills, minimizing stormwater runoff, promoting healthier gardens, and fostering environmental stewardship.

3. A well-designed rainwater harvesting system consists of several key components, including catchment surface, gutters and downspouts, first flush diverter, storage tank, filtration system, and distribution system.
4. To implement a rainwater harvesting system, assess your water needs, analyze rainfall patterns, choose a suitable location, design your system, obtain permits and approvals, install the system, test and adjust, and educate your household.
5. Regular maintenance of your rainwater harvesting system is crucial for optimal performance, including inspecting and cleaning gutters and downspouts, checking the roof and catchment surface, monitoring the pre-filtration system, inspecting storage tanks and connections, testing water quality, and maintaining pumps and treatment systems.
6. Troubleshooting common issues, such as reduced water flow, water quality issues, and leaks or overflow can help prevent further damage and ensure the continued efficiency of your system.
7. By following a regular maintenance schedule and addressing any issues promptly, you can ensure the longevity and efficiency of your rainwater harvesting system, contributing to a greener and more resilient future for our planet.
8. Embracing sustainable water management at home through rainwater harvesting benefits your household, sets a positive example for your community, and inspires others to adopt eco-friendly practices.

3

ASSESSING YOUR HOME'S RAINWATER HARVESTING POTENTIAL

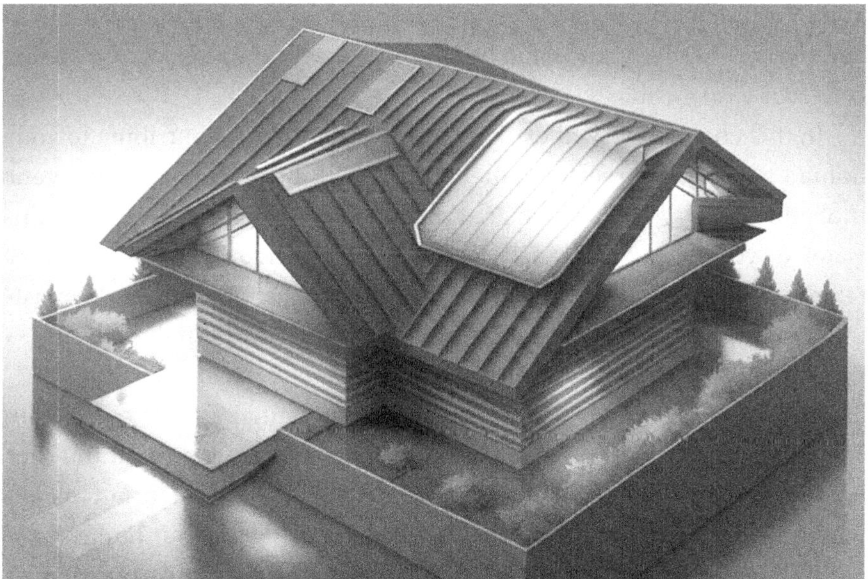

A futuristic, sleek roof and gutter system that uses advanced technology to automatically detect and redirect rainwater.

Imagine a world where you could reduce your water bill, contribute to a sustainable environment, and have a reliable source of water for your home, all at the same time. Sounds too good to be true? Well, rainwater harvesting is a practical and eco-friendly solution that can make this dream a reality. But before you jump into this exciting venture, assessing your home's rainwater harvesting potential is ***essential.*** This chapter will guide you through evaluating your home's suitability for rainwater harvesting and determining the feasibility of implementing such a system.

Rainwater harvesting is collecting, storing, and using rainwater for various purposes, such as irrigation, flushing toilets, and even drinking, after proper treatment. This ancient practice has gained renewed interest in recent years due to increasing water scarcity, rising water costs, and a growing awareness of the need for sustainable living. However, not all homes are equally suited for rainwater harvesting, and it's crucial to determine whether your home has the necessary conditions to make this endeavor worthwhile.

In this chapter, we will explore various factors that contribute to your home's rainwater harvesting potential. We will begin by evaluating your roof and gutter system, which are crucial in capturing and directing rainwater to your storage system. Next, we will analyze your home's water usage and needs to determine the potential benefits of rainwater harvesting. Finally, we will also delve into local climate and rainfall patterns, as they significantly impact the amount of rainwater you can collect.

Legal and regulatory factors are another essential aspect to consider, as some regions have specific laws and regulations regarding rainwater harvesting. Finally, we will wrap up the chapter with a conclusion to help you determine the feasibility of rainwater harvesting for your home.

By the end of this chapter, you will clearly understand your home's rainwater harvesting potential and the necessary steps to take if you decide to pursue this sustainable and cost-effective solution. So, let's embark on this exciting journey and discover the possibilities rainwater harvesting can offer for your home and the environment.

Evaluating Your Roof and Gutter System

Before diving into the world of rainwater harvesting, it's essential to evaluate the current state of your roof and gutter system. This section will guide you through assessing your home's existing infrastructure to determine if it's suitable for rainwater collection and what modifications may be necessary.

First, let's discuss the importance of having a well-maintained roof and gutter system. Your roof is the primary catchment area for rainwater. At the same time, the gutters and downspouts direct the water to your storage tanks. A poorly maintained or damaged roof can lead to water contamination, leaks, and structural issues, which can negatively impact the efficiency and safety of your rainwater harvesting system.

To begin your evaluation, inspect the material and condition of your roof. The ideal roofing material for rainwater harvesting is metal, as it's durable, easy to clean, and less likely to leach contaminants into the water. Other suitable materials include tile, slate, and asphalt shingles. However, suppose your roof is made of wood, clay, or asbestos. In that case, it's not recommended for rainwater collection due to the potential for contamination or structural issues.

Next, examine the overall condition of your roof. Look for signs of damage, such as missing or broken shingles, rust, or visible leaks. If you find any issues, **addressing them before installing a rainwater harvesting system is crucial.** Additionally, ensure that your roof is free of debris, such as leaves, branches, or moss, as these can obstruct water flow and lead to contamination.

Now, let's turn our attention to the gutter system. Gutters and downspouts are vital in directing rainwater from your roof to your storage tanks. To assess their suitability for rainwater harvesting, check for proper installation, adequate size, and good condition. Gutters should be securely fastened to your home without visible gaps or leaks. Downspouts should be large enough to handle the volume of water your roof collects during a rainstorm and should direct water away from your home's foundation.

It's also essential to keep your gutters clean and free of debris.

Clogged gutters can cause water to overflow, leading to water damage and reducing the amount of rainwater you can collect. Regularly cleaning your gutters and installing gutter guards can help prevent these issues.

Lastly, consider the layout of your gutter system and the location of your downspouts. Downspouts should be positioned near the intended location of your storage tanks to minimize the distance water must travel and reduce the potential for contamination. You can modify your gutter system or add downspouts to optimize your rainwater collection if necessary.

In conclusion, evaluating your roof and gutter system is crucial in determining the feasibility of rainwater harvesting for your home. Ensure your roof is made of suitable material, in good condition, and free of debris. By maintaining a well-functioning gutter system, you'll be well on your way to reaping the benefits of this sustainable and cost-effective water source.

Analyzing Your Home's Water Usage and Needs

Before diving into the world of rainwater harvesting, it's essential to have a clear understanding of your home's water usage and needs. This will help you determine the appropriate size and design of your rainwater harvesting system and ensure that it meets your household's requirements effectively. This section will discuss various aspects of analyzing your home's water usage and needs, which will ultimately help you make informed decisions about your rainwater harvesting project.

To begin with, start by examining your **water bills** for the past year. This will give you a rough estimate of your household's monthly water consumption. Remember that water usage may vary seasonally, with higher consumption during the warmer months due to gardening, lawn maintenance, and filling swimming pools. Please note these fluctuations, as they will play a crucial role in determining the capacity of your rainwater harvesting system.

Next, consider the specific water needs of your household. List all the activities and appliances that require water, such as drinking, cooking, bathing, laundry, and irrigation. It's important to remember that not all harvested rainwater needs to be treated to potable standards. For instance, water used for irrigation or toilet flushing can be lower quality than water used for drinking or cooking. By identifying the various uses of water in your home, you can design a rainwater harvesting system that caters to these specific needs, ultimately saving you money on water treatment costs.

Now that you clearly understand your home's water usage, it's time to evaluate the potential of rainwater harvesting to meet these needs. Start by calculating the percentage of your household's water usage that can be realistically replaced with harvested rainwater. This will depend on factors such as the size of your roof, the efficiency of your gutter system, and local rainfall patterns. Remember that replacing 100% of your water needs with rainwater may not be feasible, especially in areas with low rainfall or strict regulations regarding rainwater use.

Finally, consider the financial aspect of implementing a rainwater harvesting system. While the initial investment may seem high, weighing this against the long-term savings on your water bills is essential. Additionally, local incentives or rebates may be available for homeowners who invest in rainwater harvesting systems, further reducing the overall cost.

In conclusion, analyzing your home's water usage and needs is critical in determining the feasibility of rainwater harvesting for your household. By carefully considering your water consumption patterns, specific water needs, and the potential for rainwater harvesting to meet these needs, you can make informed decisions about the size and design of your rainwater harvesting system. This will help you maximize the benefits of this sustainable and environmentally-friendly practice while reducing your reliance on municipal water supplies and lowering your water bills.

Identifying Local Climate and Rainfall Patterns

Understanding your area's local climate and rainfall patterns is a crucial step in assessing your home's rainwater harvesting potential. By gaining insight into the amount and frequency of rainfall, you can better determine the size and type of rainwater harvesting system that will be most effective for your home. This section will discuss the importance of analyzing climate data, how to access reliable information, and how to use this knowledge to optimize your rainwater harvesting system.

First, let's discuss why it's essential to understand your local climate and rainfall patterns. The amount of rainwater you can collect depends on the frequency and intensity of rainfall in your area. A minor rainwater harvesting system may suffice in regions with abundant rainfall. In contrast, a more extensive system may be necessary to meet your water needs in areas with less frequent or lower-intensity rainfall. Additionally, knowing the seasonal variations in rainfall can help you plan for periods of water scarcity or surplus.

You can consult various sources such as local meteorological departments, government agencies, or online weather databases to access accurate and up-to-date climate data. These resources typically provide historical rainfall data, which can give you a good idea of your area's average annual and monthly precipitation levels. Therefore, it's essential to use reliable sources to ensure that your rainwater harvesting system is designed based on accurate information.

Once you have gathered the necessary climate data, you can **analyze the information** to determine the best rainwater harvesting system for your home. Start by calculating the average monthly rainfall and identifying any seasonal patterns. For example, suppose you live in an area with a distinct rainy season. In that case, you may need to invest in a larger storage tank to collect and store water during the drier months. On the other hand, if rainfall is evenly distributed throughout the year, a smaller tank may be sufficient.

It's also important to consider the intensity of rainfall events. For example, suppose your area experiences heavy downpours. In that case, your rainwater harvesting system should be designed to handle the increased flow rate, which may require more oversized gutters and downspouts. Conversely, a system with smaller components may be adequate if your region has more frequent, lighter rainfall.

In conclusion, identifying local climate and rainfall patterns is critical to assessing your home's rainwater harvesting potential. By understanding the amount, frequency, and intensity of rainfall in your area, you can design a system tailored to your specific needs and optimizes the collection and storage of this valuable resource. Remember to consult reliable sources for climate data and consider seasonal variations when planning your rainwater harvesting system.

Considering Legal and Regulatory Factors

As you embark on the journey to harness the power of rainwater for your home, it is crucial to consider the legal and regulatory factors that may impact your rainwater harvesting project. While rainwater harvesting is generally encouraged as an environmentally friendly and sustainable practice, specific rules and regulations within your local area may govern its implementation. This section will discuss the importance of understanding these legal and regulatory factors and guide how to navigate them.

First and foremost, it is essential to familiarize yourself with any local, state, or federal laws and regulations that pertain to rainwater harvesting. These laws can vary widely depending on your location, so it is essential to **research and understand** the specific requirements for your area. For example, some regions may have strict guidelines on the size and type of rainwater collection systems allowed. In contrast, others might offer incentives for implementing such systems. In some cases, permits may be required to install rainwater harvesting systems, especially if they involve significant alterations to your home's structure or plumbing.

In addition to legal requirements, you must consider any regulations or guidelines set forth by your homeowners' association (HOA) or other local governing bodies. These organizations may have specific rules regarding the appearance and placement of rainwater collection systems, mainly if they are visible from the street or neighboring properties. Review relevant documents, such as your HOA's bylaws or covenants, conditions, and restrictions (CC&Rs), to ensure your planned rainwater harvesting system complies with these rules.

Another critical aspect to consider is the potential impact of your rainwater harvesting system on your neighbors and the surrounding environment. While rainwater collection can provide numerous benefits, it is essential to ensure that your system does not cause any harm or inconvenience to others. For example, ensure your system does not divert water from neighboring properties or create standing water that could attract pests or create safety hazards. If you need clarification on your system's potential impacts, consult a professional to help you design and implement a reliable and effective rainwater harvesting solution.

Finally, staying informed about any changes to legal and regulatory requirements that may affect your rainwater harvesting system is essential. Laws and regulations can evolve, and staying up-to-date on these changes can help you maintain compliance and avoid any potential issues. Consider subscribing to newsletters or joining local organizations focused on sustainability and water conservation to stay informed about relevant updates.

In conclusion, understanding and adhering to the legal and regulatory factors surrounding rainwater harvesting is a crucial step in ensuring the success of your project. By taking the time to research and comply with these requirements, you can confidently move forward with your rainwater harvesting journey, knowing that your system is both environmentally responsible and legally sound.

Determining the Feasibility of Rainwater Harvesting for Your Home

As we reach the end of this chapter, it's time to take a step back and consider the bigger picture. By now, you should understand the various factors contributing to your home's rainwater harvesting potential. From evaluating your roof and gutter system to analyzing your home's water usage and needs, you've gathered valuable information that will help you decide whether rainwater harvesting is a viable option for your household.

In addition to the practical aspects of rainwater harvesting, it's essential to consider your area's local **climate and rainfall patterns.** After all, the success of your rainwater harvesting system will largely depend on the amount and frequency of rainfall your region experiences. By identifying these patterns, you can better estimate the potential benefits of implementing a rainwater harvesting system and adjust your design as needed.

Of course, knowing any legal and regulatory factors that may impact your ability to harvest rainwater is crucial. For example, while many regions encourage rainwater harvesting as a sustainable practice, others may have restrictions or guidelines that you'll need to follow. By familiarizing yourself with these regulations, you can ensure your system is compliant and avoid potential issues.

So, is rainwater harvesting feasible for your home? Ultimately, the answer will depend on your unique circumstances and the information you've gathered throughout this chapter. Suppose, for example, suppose you've discovered that your roof and gutter system are well-suited for rainwater collection. In that case, your household water usage aligns with the potential benefits of a rainwater harvesting system. Furthermore, if your local climate and regulations are favorable, rainwater harvesting could be a valuable addition to your home.

However, if you've identified significant challenges or limitations in these areas, weighing the potential benefits against the costs and effort required to overcome these obstacles is essential. In some cases, exploring alternative water-saving strategies or modifying your home to improve its rainwater harvesting potential may be more practical.

Regardless of your final decision, assessing your home's rainwater harvesting potential has undoubtedly provided you with valuable insights into your household's water usage and the environmental factors at play. By taking the time to understand these aspects, you're better equipped to make informed choices about your home's water management and contribute to a more sustainable future.

Chapter Summary

1. Rainwater harvesting is a practical and eco-friendly solution that can reduce water bills, contribute to a sustainable environment, and provide a reliable water source for your home.

2. Evaluating your roof and gutter system is crucial for determining the feasibility of rainwater harvesting, as they play a significant role in capturing and directing rainwater to your storage system.
3. Analyzing your home's water usage and needs helps determine the appropriate size and design of your rainwater harvesting system and ensures it effectively meets your household's requirements.
4. Understanding local climate and rainfall patterns is essential for designing a rainwater harvesting system tailored to your specific needs and optimizing the collection and storage of rainwater.
5. Familiarizing yourself with legal and regulatory factors surrounding rainwater harvesting is essential to ensure your system is compliant and avoids potential issues.
6. Considering the potential impact of your rainwater harvesting system on neighbors and the surrounding environment is essential for responsible and effective implementation.
7. Staying informed about legal and regulatory requirements changes can help you maintain compliance and adapt your rainwater harvesting system as needed.
8. Determining the feasibility of rainwater harvesting for your home depends on your unique circumstances, including your roof and gutter system, household water usage, local climate, and regulations.

4

DESIGNING AND PLANNING YOUR RAINWATER HARVESTING SYSTEM

A green, eco-friendly house with a rainwater system installation that collects rainwater from the roof.

Rainwater harvesting is an ancient practice that has renewed interest recently as homeowners seek sustainable and eco-friendly solutions to reduce their water consumption and reliance on municipal water supplies. A well-designed rainwater harvesting system can provide a **valuable source of water** for your home and contribute to the conservation of water resources and the reduction of stormwater runoff. This chapter will guide you through the essential steps of designing and planning a rainwater harvesting system tailored to your home's unique needs and characteristics.

Designing and planning a rainwater harvesting system may seem daunting. Still, it can be an enjoyable and rewarding experience with the right approach and guidance. The key to a successful rainwater harvesting system is careful planning, considering your home's specific requirements and the local climate conditions. By following the steps outlined in this chapter, you will be able to create a system that is efficient and effective in meeting your water needs while also being aesthetically pleasing and easy to maintain.

In the following sections, we will explore the various aspects of designing and planning a rainwater harvesting system, including assessing your home's rainwater collection potential, choosing the right components, creating a maintenance and monitoring plan, and integrating the system with your home's existing infrastructure. By the end of this chapter, you will clearly understand the essential elements of a rainwater harvesting system and be well-equipped to bring your vision to life.

So, let's embark on this exciting journey towards a more sustainable and water-efficient home by diving into the world of rainwater harvesting system design and planning.

Assessing Your Home's Rainwater Collection Potential

Before diving into the rainwater harvesting world, evaluating your home's rainwater collection potential is essential. This assessment will help you determine the feasibility of installing a rainwater harvesting system and guide you in selecting the right components to maximize your water savings. This section will walk you through the key factors to consider when assessing your home's rainwater collection potential.

Roof Area: The size of your roof plays a significant role in determining the amount of rainwater you can collect. Generally, the larger the roof area, the more rainwater you can harvest. To calculate your roof's catchment area, measure the length and width of your roof and multiply them together. Remember that roofs with steeper slopes may have a larger catchment area than flat roofs.

Rainfall Patterns: The amount of rainfall in your region is another crucial factor. Research the average annual rainfall in your area and note any seasonal variations. This information will help you estimate the volume of rainwater you can collect and decide on the appropriate storage capacity for your system.

Roof Material: The type of roofing material on your home can affect the quality and quantity of rainwater you can collect. Metal roofs, for example, are ideal for rainwater harvesting as they are smooth and allow for efficient water flow. On the other hand, roofs made of asphalt shingles or clay tiles may require additional filtration systems to ensure the collected water is clean and safe for use.

Gutter and Downspout System: Your home's gutter and downspout system is crucial in directing rainwater from your roof to your storage tank. Inspect your gutters and downspouts for any signs of damage or blockages that may hinder water flow. Additionally, consider the size and layout of your gutter system to ensure it can handle the rainwater you plan to collect.

Available Space for Storage: The space for installing a storage tank is another critical consideration. Assess your property to identify suitable locations for a rainwater storage tank, considering factors such as accessibility for maintenance, distance from your home's foundation, and any local regulations or zoning requirements.

Intended Use of Collected Rainwater: Consider how you plan to use the harvested rainwater. Suppose you intend to use it for non-potable purposes, such as irrigation or toilet flushing. In that case, you may not require extensive filtration or treatment systems. However, suppose you plan to use the collected rainwater for drinking or cooking. In that case, you must invest in a more advanced purification system to ensure the water is safe for consumption.

By carefully assessing your home's rainwater collection potential, you can make informed decisions about the design and components of your rainwater harvesting system. This evaluation will help you maximize your water savings and ensure that your system is tailored to your specific needs and circumstances.

Choosing the Right Rainwater Harvesting Components

Selecting the appropriate components for your rainwater harvesting system is crucial to its efficiency and longevity. You can create a practical and aesthetically pleasing system by understanding the various options available and how they fit into your overall design. This section will discuss the critical components of a rainwater harvesting system and guide how to choose the best options for your home.

Catchment Surface: The catchment surface is where rainwater is collected, typically on your home's roof. The material and slope of your roof will impact the quality and quantity of water collected. Metal roofs are ideal for rainwater harvesting, as they are durable, low-maintenance, and provide a smooth surface for water to flow. Asphalt shingles, while common, may leach chemicals into the water and require additional filtration. Consider your roof's material and condition when planning your system.

Gutters and Downspouts: Gutters and downspouts are essential for directing water from your roof to your storage tank. Choose durable, corrosion-resistant materials, and easy to clean, such as aluminum or stainless steel. Ensure your gutters are appropriately sized to handle the water your roof collects. In addition, install leaf guards to prevent debris from entering your system.

Pre-filtration: Pre-filtration is essential in maintaining water quality and preventing debris from entering your storage tank. Various pre-filtration options include leaf screens, first-flush diverters, and roof washers. Leaf screens prevent large debris from entering your downspouts. At the same time, first flush diverters and roof washers remove the initial dirty water from your roof, ensuring only clean water enters your storage tank.

Storage Tank: The storage tank is the heart of your rainwater harvesting system, and choosing the right size and material is crucial. Your tank should be large enough to store water during heavy rainfall and provide a sufficient supply during dry spells. Consider your home's water usage, local rainfall patterns, and available space when selecting a tank size. Tanks can be made from various materials, including polyethylene, fiberglass, and stainless steel. Each material has pros and cons, so research and compare options before deciding.

Pump and Pressure System: To utilize the harvested rainwater throughout your home, you will need a pump and pressure system. Select an energy-efficient and powerful pump to meet your home's water pressure requirements. Additionally, consider installing a pressure tank to maintain consistent water pressure and reduce the frequency of pump cycling.

Filtration and Disinfection: To ensure the harvested rainwater is safe, it must be adequately filtered and disinfected. Filtration options include sediment filters, activated carbon filters, and ultraviolet (UV) filters, which remove particles, chemicals, and pathogens. Depending on your water quality and intended use, you may need a combination of these filters. For disinfection, UV light or chlorination systems can kill any remaining bacteria or viruses.

Overflow and Drainage: Finally, your rainwater harvesting system should have an overflow mechanism to prevent your storage tank from overfilling during heavy rainfall. This overflowing water can be directed to a garden, stormwater drain, or another appropriate location.

By carefully selecting the right components for your rainwater harvesting system, you can create a reliable, efficient, and environmentally friendly water source for your home. Consider your home's unique needs and local regulations when choosing components, and consult with professionals if necessary. With proper planning and execution, your rainwater harvesting system will be a valuable addition to your home for years.

Creating a Maintenance and Monitoring Plan

A well-designed rainwater harvesting system is a valuable addition to your home. Still, like any other infrastructure, it requires regular maintenance and monitoring to ensure optimal performance and longevity. In this section, we will discuss the importance of creating a comprehensive maintenance and monitoring plan and the key steps to include in your plan to keep your rainwater harvesting system running smoothly.

Regular maintenance of your rainwater harvesting system is essential for several reasons:

1. It helps to prevent the buildup of debris, algae, and other contaminants that can affect the quality of the harvested rainwater.
2. It ensures that all system components, such as gutters, filters, and storage tanks, function correctly and efficiently.
3. A well-maintained system can reduce the risk of leaks, water damage, and other issues that can lead to costly repairs or even compromise the structural integrity of your home.

To create an effective maintenance and monitoring plan, consider the following steps:

Schedule regular inspections: Set a schedule for inspecting your rainwater harvesting system, including gutters, downspouts, filters, storage tanks, and any other components. This can be done monthly, quarterly, or annually, depending on your system's complexity and the local climate. During these inspections, check for signs of wear and tear, leaks, or damage, and make any necessary repairs or replacements.

Clean and maintain gutters and downspouts: Regularly clean your gutters and downspouts to prevent the buildup of leaves, dirt, and other debris that can impede the flow of rainwater and lead to clogs or damage. This is especially important during the fall and spring seasons, when debris accumulation is more likely.

Monitor and clean filters: Check your rainwater filters regularly to ensure they are not clogged or damaged. Clean or replace filters as needed to maintain the quality of the harvested water and prevent strain on your system's components.

Inspect and maintain storage tanks: Regularly inspect your storage tanks for signs of leaks, corrosion, or other damage. Clean the interior of the tanks periodically to prevent the growth of algae and bacteria and ensure that the tank's vents and overflow systems are functioning correctly.

Monitor water quality: Regularly test the quality of your harvested rainwater to ensure it meets the appropriate standards for its intended use, such as irrigation, toilet flushing, or even potable water if adequately treated. This can help you identify issues with your system's filtration or treatment processes and make necessary adjustments.

Keep a maintenance log: Document all maintenance and monitoring activities, including dates, findings, and any repairs or replacements. This log can serve as a valuable reference for future maintenance. In addition, it can help you identify patterns or recurring issues requiring further attention.

Stay informed about local regulations and best practices: Regularly review local regulations and guidelines for rainwater harvesting to ensure your system remains compliant. Additionally, stay informed about new technologies, best practices, and other advancements in rainwater harvesting to optimize your system's performance and efficiency.

By following these steps and creating a comprehensive maintenance and monitoring plan, you can ensure the longevity and effectiveness of your rainwater harvesting system. This will help you conserve water and reduce your environmental footprint and save money on water bills and potential repair costs in the long run.

Integrating Your Rainwater Harvesting System with Your Home's Existing Infrastructure

As you embark on the exciting journey of designing and planning your rainwater harvesting system, it is essential to consider how it will seamlessly integrate with your home's existing infrastructure. This integration not only ensures the efficiency and effectiveness of your rainwater harvesting system but also maintains your home's aesthetic appeal and functionality. This section will discuss various aspects of integrating your rainwater harvesting system with your home's existing infrastructure, including plumbing, electrical, and landscaping considerations.

First and foremost, evaluating your **home's plumbing system** is crucial. If adequately treated, your rainwater harvesting system must connect to your existing plumbing to supply water for various household uses, such as irrigation, toilet flushing, and even potable water. Consult a professional plumber to determine the best approach for connecting your rainwater harvesting system to your home's plumbing. This may involve installing additional pipes, rerouting existing pipes, or integrating a pump system to ensure adequate water pressure.

Next, consider the **electrical requirements** of your rainwater harvesting system. Some components, such as pumps and control systems, may require electrical connections. Working with a licensed electrician is essential to ensure that your system is safely and correctly wired, adhering to all local codes and regulations. Additionally, you plan to incorporate a solar-powered pump or other renewable energy sources. In that case, you must consider the appropriate electrical connections and equipment for these components.

Landscaping is another critical aspect of integrating your rainwater harvesting system with your home's infrastructure. Thoughtful landscaping can enhance the visual appeal of your rainwater harvesting system and improve its efficiency. For instance, strategically placed plants can help direct rainwater toward your catchment area. At the same time, porous paving materials can facilitate water infiltration into the ground, reducing runoff. Furthermore, consider incorporating rain gardens or swales into your landscape design to help manage excess rainwater and prevent flooding.

Lastly, ensuring that your rainwater harvesting system complies with all local building codes, regulations, and permitting requirements is essential. This may involve obtaining permits, working with a licensed contractor, or submitting plans for review by local authorities. Compliance with these regulations ensures the safety and effectiveness of your rainwater harvesting system and protects your investment in your home.

In conclusion, integrating your rainwater harvesting system with your home's existing infrastructure is crucial in designing and planning your system. By carefully considering plumbing, electrical, and landscaping aspects, as well as adhering to local codes and regulations, you can create a rainwater harvesting system that is both functional and aesthetically pleasing. With thoughtful integration, your rainwater harvesting system will become a valuable addition to your home, providing a sustainable water source for years.

Bringing Your Rainwater Harvesting System to Life

In conclusion, designing and planning a rainwater harvesting system for your home is exciting and rewarding. By following the steps outlined in this chapter, you can create a sustainable and efficient system that benefits the environment and reduces your water bills and reliance on municipal water supplies.

As you bring your rainwater harvesting system to life, remember that the key to success lies in carefully planning, selecting the right components, and integrating the system seamlessly with your home's infrastructure. Doing so ensures that your system operates at peak performance and provides you with a **reliable water source for years.**

Remember the importance of creating a maintenance and monitoring plan to keep your system in top shape. Regular maintenance will help you avoid potential issues and ensure that your rainwater harvesting system remains valuable to your home.

Lastly, remember that rainwater harvesting is about more than just its practical benefits. It is also an opportunity to contribute to a more sustainable future and become a responsible steward of our planet's precious resources. By harvesting rainwater, you are taking a significant step towards conserving water, reducing your carbon footprint, and promoting environmental awareness within your community.

So, as you embark on this journey to bring your rainwater harvesting system to life, take pride in knowing that you are making a positive impact on the world around you. With careful planning, dedication, and a little creativity, you can create a functional and environmentally friendly system, providing you and your family with a sustainable source of water for years to come.

Chapter Summary

1. Rainwater harvesting is an ancient practice that has gained renewed interest due to its sustainability and eco-friendliness, helping homeowners reduce water consumption and reliance on municipal water supplies.
2. Assessing your home's rainwater collection potential is crucial, considering factors such as roof area, rainfall patterns, roof material, gutter and downspout systems, available space for storage, and intended use of collected rainwater.
3. Choosing the right components for your rainwater harvesting system is essential for its efficiency and longevity, including catchment surfaces, gutters and downspouts, pre-filtration, storage tanks, pumps, pressure systems, filtration and disinfection, and overflow and drainage systems.
4. Creating a comprehensive maintenance and monitoring plan is vital for ensuring optimal performance and longevity of your rainwater harvesting system, including regular inspections, cleaning and maintaining gutters and downspouts, monitoring and cleaning filters, inspecting and maintaining storage tanks, monitoring water quality, and staying informed about local regulations and best practices.
5. Integrating your rainwater harvesting system with your home's infrastructure is crucial for seamless functionality and aesthetic appeal, considering plumbing, electrical, and landscaping aspects.
6. Compliance with local building codes, regulations, and permitting requirements is essential for the safety and effectiveness of your rainwater harvesting system and for protecting your investment in your home.

7. Thoughtful landscaping can enhance the visual appeal of your rainwater harvesting system and improve its efficiency, incorporating elements such as strategically placed plants, permeable paving materials, rain gardens, or swales.
8. Designing and planning a rainwater harvesting system is an exciting and rewarding endeavor that contributes to a more sustainable future, conserves water, reduces carbon footprint, and promotes environmental awareness within your community.

5

CHOOSING THE RIGHT RAINWATER COLLECTION AND STORAGE EQUIPMENT

A green, eco-friendly house with a rainwater system installation that collects rainwater from the roof.

Rainwater harvesting has been vital for centuries, offering numerous benefits such as reducing water bills, conserving precious resources, and promoting a sustainable lifestyle. With the growing awareness of environmental issues and the need to conserve water, many homeowners now consider rainwater harvesting an essential part of their home management. This chapter will explore the various rainwater collection and storage equipment available and guide you in making the right choices for your home's rainwater harvesting system.

Rainwater collection and storage equipment are **essential** for any rainwater harvesting system. With proper treatment, these systems are designed to capture, store, and distribute rainwater for various household uses, such as watering plants, flushing toilets, and even drinking. By investing in the right equipment, you can maximize the efficiency of your rainwater harvesting system and enjoy the numerous benefits it offers.

The process of rainwater harvesting begins with the collection of rainwater from surfaces such as roofs, gutters, and downspouts. This water is then channeled through a series of filters and storage tanks before being distributed for use. The choice of equipment for each stage of this process plays a crucial role in determining your rainwater harvesting system's overall efficiency and effectiveness.

This chapter will discuss the various types of rainwater collection systems, storage solutions, and critical factors to consider when choosing the right equipment for your home. We will also provide practical tips and insights to help you make informed decisions and ensure your rainwater harvesting system is tailored to your needs and preferences.

By the end of this chapter, you will have a comprehensive understanding of the different rainwater collection and storage equipment available and be well-equipped to make the right choices for your home's rainwater harvesting system. So, let's dive in and explore the fascinating world of rainwater harvesting equipment!

Assessing Your Home's Rainwater Harvesting Potential

Before diving into the world of rainwater harvesting, evaluating your home's potential for collecting and utilizing rainwater effectively is essential. This assessment will understand your home's unique characteristics and help you make informed decisions when selecting the proper collection and storage equipment. This section will discuss the key factors to consider when assessing your home's rainwater harvesting potential.

Roof Size and Material

The first step in assessing your home's rainwater harvesting potential is to evaluate the size and material of your roof. The roof serves as the primary catchment area for rainwater, and its size directly impacts the volume of water you can collect. Measure the square footage of your roof to estimate the amount of rainwater you can harvest.

Additionally, consider the material of your roof, as this will affect the quality of the collected rainwater. For example, metal roofs are ideal for rainwater harvesting, as they are less likely to contaminate water. On the other hand, roofs made of asphalt shingles or treated wood may release chemicals that could compromise the quality of the harvested rainwater. Finally, suppose your roof is made of something other than a suitable material. In that case, consider alternative catchment surfaces or invest in a pre-filtration system.

Rainfall Patterns and Climate

Your region's rainfall will significantly impact your home's rainwater harvesting potential. First, research the average annual rainfall in your area and note any seasonal patterns. This information will help you determine the size and capacity of your rainwater collection and storage system. A smaller storage tank may suffice in areas with consistent rainfall throughout the year. However, a larger storage capacity will be necessary to ensure an adequate water supply during dry periods in regions with distinct wet and dry seasons.

Water Usage and Demand

Consider your household's water usage and demand to design a rainwater harvesting system effectively. Analyze your water bills to determine your average monthly water consumption. Next, identify the activities and appliances that use the most water, such as gardening, laundry, and toilet flushing. This information will help you estimate the percentage of your water needs that can be met through rainwater harvesting and guide you in selecting the appropriate equipment.

Local Regulations and Restrictions

Before implementing a rainwater harvesting system, it's crucial to familiarize yourself with any local regulations and restrictions that may apply. For example, some regions may have specific guidelines on the types of collection systems allowed or require permits for installing storage tanks. Additionally, there may be restrictions on harvesting rainwater, particularly for potable purposes. Consult your local authorities and adhere to regulations to ensure a safe and compliant rainwater harvesting system.

By carefully assessing your home's rainwater harvesting potential, you can make informed decisions when selecting the proper collection and storage equipment. The following section will explore the different types of rainwater collection systems and their respective advantages and disadvantages.

Comparing Different Types of Rainwater Collection Systems

Rainwater harvesting is an environmentally friendly and cost-effective way to utilize a natural resource that often goes to waste. However, not all rainwater collection systems are created equal. This section will explore the various types of rainwater collection systems available, their advantages and disadvantages, and how to determine which one is best suited for your home.

Roof Catchment Systems

The most common type of rainwater collection system is the roof catchment system. As the name suggests, this system collects rainwater from your home's roof. The water is channeled through gutters and downspouts, then directed into a storage tank or cistern. Roof catchment systems are popular because they are relatively easy to install and can be adapted to fit most homes.

Advantages:

- Easy to install and maintain
- It can be adapted to most homes
- Efficient in collecting large volumes of water

Disadvantages:

- Water quality may be affected by roofing materials and debris
- Requires regular cleaning and maintenance of gutters and downspouts

Surface Catchment Systems

Surface catchment systems collect rainwater from ground-level surfaces, such as driveways, patios, and other paved areas. These systems typically use channels, grates, and pipes to direct water into a storage tank or cistern. Surface catchment systems are ideal for homes with limited roof space or those looking to maximize their rainwater harvesting potential.

Advantages:

- Can collect water from multiple sources
- Ideal for homes with limited roof space

Disadvantages:

- It may require more extensive installation and maintenance
- Water quality may be affected by surface contaminants

Rain Gardens and Bioswales

Rain gardens and bioswales are landscape features that capture and filter rainwater runoff. These systems use plants and soil to slow water flow, allowing it to infiltrate the ground and recharge groundwater supplies. While less efficient at collecting water for immediate use, rain gardens and bioswales are excellent options for homeowners looking to improve their property's overall water management and reduce stormwater runoff.

Advantages:

- Improves overall water management on your property
- Reduces stormwater runoff and associated pollution
- Enhances the aesthetic appeal of your landscape

Disadvantages:

- Not as efficient for immediate water collection and use
- Requires careful planning and maintenance to ensure proper function

When comparing these different types of rainwater collection systems, it's essential to consider your home's unique characteristics, such as roof size, available space, and water usage needs. By carefully evaluating your options and weighing the advantages and disadvantages of each system, you can make an informed decision that will help you maximize your rainwater harvesting potential and contribute to a more sustainable future.

Selecting the Ideal Storage Solutions for Your Needs

Now that you better understand the various rainwater collection systems available, it's time to delve into the crucial storage aspect. Choosing the proper storage solution for your harvested rainwater is essential to ensure an efficient and effective system is in place. In this section, we will explore the different storage options available, discuss their pros and cons, and provide guidance on selecting the ideal solution for your needs.

Types of Rainwater Storage Solutions

Several types of storage solutions are available for harvested rainwater, each with its own advantages and disadvantages. Some of the most common options include:

Above-ground tanks: These are the most common and affordable type of storage solution. They are easy to install and maintain and can be made from plastic, metal, or concrete. However, they can be less aesthetically pleasing and may take up valuable yard space.

Underground tanks are installed underground, making them less visible and more space-efficient. They are also less susceptible to temperature fluctuations, which can help maintain water quality. However, they can be more expensive to install and may require additional maintenance.

Rain barrels are small, above-ground containers that collect rainwater from downspouts. They are affordable and straightforward for those with limited space or budget. However, they have a smaller storage capacity and may require more frequent emptying.

Modular storage systems: These systems consist of multiple small tanks connected, allowing for storage capacity and location flexibility. They can be installed above or below ground and quickly expanded as needed. However, they may be more expensive than single-tank solutions.

Factors to Consider When Choosing a Storage Solution

When selecting the ideal storage solution for your rainwater harvesting system, consider the following factors:

Storage capacity: Determine the amount of rainwater you expect to collect and use based on your home's harvesting potential and water usage needs. Choose a storage solution that can accommodate this volume while allowing for extra capacity in case of heavy rainfall or increased usage.

Space availability: Consider the available space in your property for installing a storage solution. Above-ground tanks and rain barrels may require more yard space, while underground tanks and modular systems can be more space-efficient.

Aesthetics: If the appearance of your storage solution is essential to you, consider options that blend well with your home's exterior or can be easily concealed, such as underground tanks or modular systems.

Budget: The cost of storage solutions can vary significantly, so consider your budget when deciding. While above-ground tanks and rain barrels are generally more affordable, investing in a more expensive option like an underground tank may be worthwhile in the long run due to its durability and lower maintenance requirements.

Maintenance: Different storage solutions may require varying maintenance levels, such as cleaning, repairs, or replacement. Consider the long-term maintenance requirements of each option and choose one that best suits your lifestyle and preferences.

Selecting the ideal storage solution for your rainwater harvesting system is crucial in ensuring its efficiency and effectiveness. By considering factors such as storage capacity, space availability, aesthetics, budget, and maintenance requirements, you can make an informed decision that best suits your needs and preferences. With the proper storage solution, you can enjoy the numerous benefits of rainwater harvesting, from conserving water resources to reducing utility bills and promoting a more sustainable lifestyle.

Key Factors to Consider When Choosing Equipment

When selecting the right equipment for your home's rainwater harvesting system, there are several key factors to consider. By considering these factors, you can ensure that you choose the most suitable and efficient equipment for your needs. This section will discuss the most important aspects to consider when selecting your rainwater collection and storage equipment.

Quality and Durability: One of the most important factors to consider when choosing equipment for your rainwater harvesting system is the quality and durability of the materials used. Investing in high-quality equipment made from durable materials will ensure that your system remains functional and practical for many years. In addition, look for products made from corrosion-resistant materials, such as stainless steel, UV-resistant plastics, or fiberglass, as these will be less likely to degrade.

Capacity: Another crucial factor to consider when selecting equipment is the capacity of the collection and storage systems. The capacity of your rainwater harvesting system should be based on your household's water usage and the average rainfall in your area. It's essential to choose equipment that can handle the volume of water you expect to collect and provide adequate storage for times when rainfall is scarce.

Ease of Installation and Maintenance: When choosing equipment for your rainwater harvesting system, it's essential to consider how easy it will be to install and maintain. Look for products with clear installation instructions that require minimal tools and expertise to set up. Additionally, consider the maintenance requirements of the equipment, such as cleaning and replacing filters, and ensure that you or a professional can efficiently perform these tasks.

Cost and Budget: Your budget will undoubtedly play a significant role in the equipment you choose for your rainwater harvesting system. While investing in high-quality, durable equipment is essential, it's also important to find products that fit your budget. Be sure to compare the prices and features of different products to find the best value for your money.

Aesthetics: While the primary purpose of your rainwater harvesting system is to collect and store water, it's also essential to consider the visual impact of the equipment on your property. Choose equipment that complements the design and style of your home, and consider options that can be easily concealed or integrated into your landscape.

Local Regulations and Requirements: Before selecting equipment for your rainwater harvesting system, research any local regulations and requirements that may apply. Some areas may have specific guidelines regarding the types of materials that can be used, the size of storage tanks, or the installation process. Ensure the equipment you choose complies with these regulations to avoid potential issues.

In conclusion, taking the time to carefully consider these key factors when choosing equipment for your home's rainwater harvesting system will help ensure you make the best choices for your needs. By selecting high-quality, durable equipment that meets your capacity requirements, is easy to install and maintain, fits within your budget, and complies with local regulations, you can create an efficient and effective rainwater harvesting system that will serve your household for years to come.

Making the Right Choices for Your Home's Rainwater Harvesting System

In conclusion, rainwater harvesting is environmentally friendly and cost-effective for meeting your home's water needs. Choosing the proper collection and storage equipment ensures that your rainwater harvesting system operates efficiently and effectively. This chapter has guided you to **make informed decisions** when selecting the appropriate equipment for your home's rainwater harvesting system.

To recap, assessing your home's rainwater harvesting potential is essential by considering factors such as your local climate, roof size, and water usage requirements. This information will help you determine the best rainwater collection system for your needs, whether a simple gutter and downspout system or a more advanced system with filters and pumps.

Regarding storage solutions, you must consider capacity, material, and location factors. Selecting the right storage tank or cistern ensures that your harvested rainwater remains clean and safe. Additionally, it is crucial to consider the key factors when choosing equipment, such as durability, ease of maintenance, and cost.

As you embark on your rainwater harvesting journey, remember that the choices you make regarding equipment will significantly impact your system's overall success and efficiency. By taking the time to research and select the best equipment for your needs, you can enjoy the numerous benefits that come with rainwater harvesting, including reduced water bills, a more sustainable lifestyle, and the satisfaction of contributing to a greener planet.

Ultimately, the right choices for your home's rainwater harvesting system will depend on your unique circumstances and preferences. By carefully considering your options and following the guidance in this chapter, you can create a customized rainwater harvesting system that meets your needs and helps you achieve your sustainability goals. Happy harvesting!

Chapter Summary

1. Rainwater harvesting is an environmentally friendly and cost-effective solution for meeting your home's water needs, offering reduced water bills and a more sustainable lifestyle.
2. Assess your home's rainwater harvesting potential by considering roof size, material, climate, water usage, and regulations.
3. There are various rainwater collection systems, including roof catchment systems, surface catchment systems, and rain gardens or bioswales. Each has advantages and disadvantages, so choose the one that best suits your needs.
4. Storage solutions for harvested rainwater include above-ground tanks, underground tanks, rain barrels, and modular storage systems. Consider storage capacity, space availability,

aesthetics, budget, and maintenance requirements when choosing the right solution.

5. When selecting equipment for your rainwater harvesting system, prioritize quality and durability to ensure long-lasting functionality and effectiveness.
6. Consider the ease of installation and maintenance when choosing equipment, as this will impact your system's overall efficiency and user-friendliness.
7. Budget and local regulations are essential when selecting equipment for your rainwater harvesting system. Ensure your chosen equipment complies with local guidelines and fits within your budget.
8. By carefully considering your options and following the guidance in this chapter, you can create a customized rainwater harvesting system that meets your needs and helps you achieve your sustainability goals.

6

INSTALLING YOUR RAINWATER HARVESTING SYSTEM

A farmer collects water from a well in the middle of his lush green field.

This chapter will delve into the exciting process of installing your rainwater harvesting system. You have learned about the numerous benefits of rainwater harvesting, the various types of systems available, and the legal aspects to consider. Then, it is time to plunge and start the installation process.

Installing a rainwater harvesting system can be a rewarding and fulfilling experience. Not only will you be contributing to a more sustainable environment, but you will also be saving money on your water bills and ensuring a reliable water source for your home. This chapter will guide you through the entire process, from assessing your home and property for rainwater harvesting to choosing the right components and following a step-by-step installation guide.

The idea of installing a rainwater harvesting system might seem ***daunting*** at first. However, with the proper guidance and a little bit of patience, you will soon be on your way to enjoying the many benefits of this eco-friendly and cost-effective solution. Our goal is to provide you with clear and concise information and an engaging tone to make this process as enjoyable and stress-free as possible.

In the following sections, we will cover the essential steps to install your rainwater harvesting system successfully. We will provide a detailed, step-by-step guide to the installation process. By the end of this chapter, you will have all the knowledge and tools necessary to install your rainwater harvesting system confidently. So, let's get started on this exciting journey toward a more sustainable and self-sufficient home!

Step-by-Step Installation Process

Now that you have assessed your home, selected the right components for your rainwater harvesting system, and familiarized yourself with the necessary maintenance and troubleshooting tips, it's time to dive into the installation process. This section will guide you through a step-by-step process to ensure your rainwater harvesting system is installed correctly and efficiently.

Prepare the site: Clear where your rainwater harvesting system will be located before installation. Remove any debris, rocks, or vegetation that may obstruct the installation process or the system's functionality. Ensure the ground is level and stable to support the tank's weight when it's full of water.

Install the gutters and downspouts: If your home still needs them, you'll need to install them. Ensure they are appropriately secured and sloped to direct water toward the downspouts. Install leaf guards or gutter screens to prevent debris from entering the system.

Set up the catchment surface: The catchment surface is where rainwater is collected, usually on your roof. Ensure that your roof is clean and free of debris. Install a roof washer to filter out contaminants before the water enters your system.

Assemble and install the tank: Follow the manufacturer's instructions to assemble your storage tank. Once assembled, place the tank on a solid, level foundation near the downspout. Ensure that the tank is adequately secured and stable.

Connect the downspout to the tank: Attach a diverter or a flexible pipe to the downspout, directing the water flow into the storage tank. Make sure all connections are secure and watertight to prevent leaks.

Install the overflow system: An overflow system is essential to prevent your tank from overflowing during heavy rainfall. Connect an overflow pipe to the designated outlet on your tank and direct the excess water away from your home's foundation.

Set up the filtration system: Install a pre-tank filter to remove debris and contaminants before the water enters your storage tank. If your system includes a post-tank filter for additional purification, install it according to the manufacturer's instructions.

Install the pump and distribution system: If your rainwater harvesting system includes a pump, follow the manufacturer's guidelines for installation. Connect the pump to your storage tank and the distribution system, which may include hoses, irrigation systems, or indoor plumbing.

Test the system: Once everything is installed, test your rainwater harvesting system by running water through the gutters, downspouts, and storage tank. Check for any leaks or issues with the filtration, overflow, and distribution systems.

Make any necessary adjustments: After testing your system, make any necessary adjustments to ensure optimal performance. This may include tightening connections, adjusting the slope of your gutters, or fine-tuning the pump's settings.

By following these steps, you'll have a fully functional rainwater harvesting system that will provide you with a sustainable and eco-friendly water source for years. The following section will discuss maintenance and troubleshooting tips to keep your system running smoothly.

Enjoying the Benefits of Your Rainwater Harvesting System

Congratulations! You have successfully installed your rainwater harvesting system and are now ready to reap its numerous benefits. As you begin to utilize this eco-friendly and cost-effective solution, you will soon realize its positive impact on your home, wallet, and environment.

First and foremost, you will notice a **significant reduction in your water bills.** By harvesting rainwater, you are tapping into an accessible and renewable water source that can be used for various purposes around your home, such as watering your garden, flushing toilets, and even doing laundry. This not only saves you money but also helps conserve precious water resources.

In addition to the financial benefits, your rainwater harvesting system will contribute to a greener and more sustainable environment. By capturing and using rainwater, you reduce your reliance on municipal water supplies, often from overexploited sources like rivers and groundwater. This helps protect these vital resources and promotes a more balanced water cycle.

Moreover, your garden and plants will thrive with the use of rainwater. Unlike treated municipal water, rainwater is free of chemicals like chlorine and fluoride, which can harm your plants. Rainwater also contains natural minerals and nutrients that promote healthy plant growth, leading to a more vibrant and flourishing garden.

Another advantage of your rainwater harvesting system is its ability to reduce stormwater runoff. Capturing rainwater prevents it from flowing over paved surfaces and picking up pollutants like oil, pesticides, and fertilizers. This helps protect local waterways and ecosystems from contamination and reduces the risk of flooding in your community.

Lastly, your rainwater harvesting system will give you a sense of accomplishment and pride. By taking this step towards a more sustainable lifestyle, you are actively preserving our planet for future generations. Furthermore, your efforts will inspire others to follow suit and contribute to a global movement toward water conservation and environmental stewardship.

In conclusion, your rainwater harvesting system is a valuable addition to your home that offers numerous benefits. As you enjoy the financial savings, environmental advantages, and the satisfaction of contributing to a greener world, you can take pride knowing that you have made a wise and responsible choice. So, go ahead and enjoy the fruits of your labor, and let your rainwater harvesting system serve as a constant reminder of your commitment to a sustainable future.

Chapter Summary

1. Installing a rainwater harvesting system is a rewarding and fulfilling experience that contributes to a more sustainable environment, saves money on water bills, and ensures a reliable water source for your home.
2. The step-by-step installation process involves preparing the site, installing gutters and downspouts, setting up the catchment surface, assembling and installing the tank, connecting the downspout to the tank, installing the overflow system, setting up the filtration system, installing the pump and distribution system, testing the system, and making any necessary adjustments.
3. The benefits of a rainwater harvesting system include reduced water bills, a greener and more sustainable environment, healthier plants and gardens, reduced stormwater runoff, and

a sense of accomplishment and pride in contributing to a sustainable future.

4. By installing and maintaining a rainwater harvesting system, homeowners can enjoy financial savings, environmental advantages, and the satisfaction of contributing to a greener world, inspiring others to follow suit and promote water conservation and environmental stewardship.

7

MAINTAINING AND MONITORING YOUR RAINWATER HARVESTING SYSTEM

A farmer collects water from a well in the middle of his lush green field.

Congratulations on taking the first step towards a more sustainable and eco-friendly lifestyle by installing a rainwater harvesting system for your home! As you reap the benefits of this environmentally responsible decision, you must understand the importance of maintaining and monitoring your system to ensure its longevity and optimal performance. This chapter will provide the necessary knowledge and tools to keep your rainwater harvesting system in excellent condition, allowing you to enjoy its benefits for years.

A well-designed and properly installed rainwater harvesting system can provide a **reliable and sustainable water source** for various household uses, such as irrigation, toilet flushing, and even drinking water, if properly treated. However, like any other system, it requires regular maintenance and monitoring to function efficiently and provide clean, safe water for your home. This chapter will guide you through the essential steps to keep your system in top shape, from regular inspection and cleaning of gutters and downspouts to monitoring water quality and system efficiency.

By following the recommendations and best practices outlined in this chapter, you will ensure the smooth operation of your rainwater harvesting system, protect your investment, and contribute to a healthier environment. So, let's dive into maintaining and monitoring your rainwater harvesting system and learn how to make the most of this valuable resource.

Regular Inspection and Cleaning of Gutters and Downspouts

A well-maintained rainwater harvesting system is crucial for ensuring optimal performance and longevity. One of the critical components in this process is the regular inspection and cleaning of gutters and downspouts. This section will discuss the importance of keeping these components clean, how often they should be inspected, and the best practices for maintaining them.

Gutters and downspouts play a vital role in the rainwater harvesting process. They collect and channel rainwater from your roof to the storage tanks, ensuring it is efficiently captured and stored for future use. However, over time, debris such as leaves, twigs, and dirt can accumulate in your gutters and downspouts, causing blockages and reducing the efficiency of your rainwater harvesting system. Therefore, **regular inspection and cleaning of these components are essential** to prevent such issues and maintain the overall performance of your system.

It is recommended that you inspect your gutters and downspouts at least twice a year, ideally during the spring and fall seasons when debris accumulation is most likely to occur. However, if you live in an area with heavy foliage or frequent storms, you may need to inspect and clean your gutters more frequently. Look for any signs of blockages, such as overflowing gutters during rainfall or water pooling around the base of your downspouts.

When it comes to cleaning your gutters and downspouts, safety should be your top priority. Use a sturdy ladder to access your gutters and ensure it is properly secured before you begin. It is also a good idea to wear gloves and safety goggles to protect yourself from any debris or insects that may be present.

Begin by removing large debris, such as leaves and twigs, from your gutters or with a gutter scoop. Next, use a garden hose with a high-pressure nozzle to remove any remaining debris and dirt from the gutters and downspouts. Finally, direct the water flow away from your home's foundation to prevent potential water damage.

While cleaning your gutters and downspouts, please take the opportunity to inspect them for any signs of damage, such as cracks, rust, or loose connections. If you notice any issues, it is essential to address them promptly to prevent further damage to your rainwater harvesting system.

In conclusion, regular inspection and cleaning of your gutters and downspouts are essential for maintaining the efficiency and longevity of your rainwater harvesting system. By keeping these components clean and in good working order, you can ensure that your system continues to provide you with a reliable source of water for years to come.

Checking and Maintaining Storage Tanks and Filters

Proper maintenance of storage tanks and filters is crucial to your rainwater harvesting system's overall performance and longevity. This section will discuss the importance of regular inspections, cleaning, and maintenance of these components to ensure that your system continues to provide clean and safe water for your household needs.

Storage tanks are the heart of your rainwater harvesting system, storing the collected water until needed. Unfortunately, over time, sediment, debris, and algae can accumulate in the tanks, affecting water quality and even damaging the tank itself. To prevent these issues, inspecting and maintaining your storage tanks regularly is essential.

Begin by visually inspecting the tank for any signs of damage, such as cracks, leaks, or corrosion. If you notice any issues, you must address them immediately to prevent further damage or contamination of your water supply. Additionally, check for any signs of algae growth, indicating inadequate sunlight protection or poor water circulation. If algae are present, you may need to clean the tank and consider adding a UV light or other algae control measures.

To clean your storage tank:

1. Drain the water and remove any sediment or debris settled at the bottom. You can use a wet-dry vacuum or a specialized tank cleaning tool to accomplish this task.
2. Scrub the tank's interior surfaces with a long-handled brush and a mild, non-toxic cleaning solution.
3. Rinse the tank thoroughly with clean water and allow it to dry before refilling.

Filters play a vital role in maintaining the quality of your harvested rainwater by removing debris, sediment, and other contaminants before the water enters your storage tank. Regular inspection and cleaning of your filters are necessary to ensure their effectiveness and prevent clogging.

First, inspect the filters for any signs of damage or wear, such as tears or holes in the mesh or filter media. If you find any issues, replace the damaged components immediately. Next, remove the filter media and clean it according to the manufacturer's instructions. This may involve rinsing, brushing, or soaking the media in a cleaning solution. Finally, replace the filter media if it shows signs of wear or if the manufacturer recommends replacement after a certain period.

Finally, check the filter housing and connections for any signs of leaks or damage. Tighten any loose connections and replace any damaged components as needed. Regularly inspecting and maintaining your storage tanks and filters ensure that your rainwater harvesting system continues to provide clean, safe water for your home while minimizing the risk of system failure or damage.

Monitoring Water Quality and System Efficiency

Ensuring the quality of the harvested rainwater and the overall efficiency of your rainwater harvesting system is crucial for the success of your sustainable water management efforts. In this section, we will discuss the importance of monitoring water quality and system efficiency and provide practical tips and guidelines to help you keep your rainwater harvesting system in optimal condition.

One of the primary reasons for monitoring water quality is to ensure that the harvested rainwater is **safe for its intended use.** Whether you plan to use the water for irrigation, household cleaning, or even for drinking, it is essential to regularly test the water quality to avoid potential health risks and maintain the effectiveness of your system.

To monitor water quality, you can start by visually inspecting the water for any signs of contamination, such as discoloration, cloudiness, or debris. Additionally, you can use water testing kits available in the market to check for specific contaminants like bacteria, heavy metals, and chemicals. These kits usually come with easy-to-follow instructions and provide quick results, allowing you to take immediate action if necessary.

Another essential aspect of monitoring your rainwater harvesting system is evaluating its efficiency. This involves assessing the performance of various components, such as gutters, downspouts, filters, and storage tanks, to ensure they function optimally. In addition, regularly inspecting these components for signs of wear, damage, or blockage can help you identify potential issues early on and address them before they escalate.

To gauge the efficiency of your system, consider tracking the amount of water you collect and use over time. This can help you identify any discrepancies between the expected and actual water yield, which may indicate issues with your system. Additionally, monitoring your water usage can help you make informed decisions about future system upgrades or adjustments to improve efficiency.

In summary, monitoring water quality and system efficiency is essential to maintaining your rainwater harvesting system. By regularly inspecting and testing the water and evaluating the performance of various system components, you can ensure the longevity and optimal performance of your rainwater harvesting system. This contributes to a more sustainable water management approach and helps you enjoy the numerous benefits of rainwater harvesting for years to come.

Troubleshooting Common Issues and Seeking Professional Help

As with any home system, your rainwater harvesting setup may occasionally encounter issues requiring troubleshooting or professional assistance. In this section, we will discuss some common problems you may face and provide guidance on when to call in the experts.

One common issue that homeowners may experience is a **decrease in water pressure.** Various factors, such as clogged filters, blocked pipes, or malfunctioning pumps, can cause this. If you notice a drop in water pressure, check your filters and clean or replace them as necessary. If the issue persists, inspect your pipes for blockages and ensure your pump functions correctly. If you cannot identify or resolve the problem, it may be time to consult a professional.

Another issue that may arise is a **leak in your storage tank or pipes.** Regularly inspect your tank and pipes for any signs of leakage, such as damp spots, mold, or pooling water. If you discover a leak, it is crucial to address it promptly to prevent water loss and potential damage to your property. Small leaks may be repairable with a waterproof sealant, but more significant leaks or cracks may necessitate professional intervention.

Water quality is a crucial aspect of your rainwater harvesting system, and any changes in taste, odor, or appearance should be taken seriously. If you notice any unusual changes in your harvested water, it is essential to test it for contaminants. You can purchase water testing kits online or from local home improvement stores. If the test results indicate the presence of harmful substances, consult a professional to determine the appropriate course of action.

Sometimes, your rainwater harvesting system **may collect less water than you anticipated.** This could be due to issues with your catchment area, such as insufficient roof surface area or poor gutter design. It could also result from leaks or evaporation from your storage tank. Suppose you suspect that your system is underperforming. In that case, consulting with a professional who can assess your setup and recommend improvements may be worthwhile.

Finally, regular maintenance is critical to preventing many of these common issues. By staying vigilant and addressing any problems as they arise, you can help ensure your rainwater harvesting system's longevity and optimal performance. However, if you ever feel unsure about handling a particular issue or if your system requires more extensive repairs, feel free to seek the help of a professional. Their expertise can save you time, money, and frustration in the long run and help you maintain a safe and efficient rainwater harvesting system for your home.

Ensuring Longevity and Optimal Performance of Your Rainwater Harvesting System

In conclusion, diligent maintenance and monitoring is the key to ensuring the longevity and optimal performance of your rainwater harvesting system. By taking the time to regularly inspect, clean, and maintain the various components of your system, you can prevent potential issues from escalating and ensure that your system continues to function efficiently.

The gutters and downspouts are crucial in directing rainwater to your storage tanks. Keeping them clean and free from debris will improve the quality of the harvested water and prevent damage to your roof and home structure. Similarly, regularly checking and maintaining your storage tanks and filters will help to preserve water quality and extend the life of your system.

Monitoring water quality and system efficiency is essential to guarantee that your rainwater harvesting system provides a reliable and safe water source. By conducting periodic tests and keeping an eye on the overall performance of your system, you can identify any issues early on and take the necessary steps to address them.

If you encounter any problems or concerns with your rainwater harvesting system, feel free to seek professional help. A qualified expert can provide valuable guidance and assistance in troubleshooting and resolving any issues that may arise.

By following the tips and advice outlined in this chapter, you can take pride in knowing that you are contributing to a more sustainable and environmentally friendly way of life and enjoying the numerous benefits of having a well-maintained and efficient rainwater harvesting system. With proper care and attention, your system will continue to serve you and your home for many years.

Chapter Summary

1. Regular maintenance and monitoring of your rainwater harvesting system are crucial for ensuring its longevity and optimal performance.
2. Inspect and clean gutters and downspouts at least twice a year to prevent blockages and maintain the efficiency of your system.

3. Regularly check storage tanks and filters for signs of damage, wear, or blockage, and clean or replace them as necessary to maintain water quality and system performance.
4. Monitor water quality by visually inspecting the water and using water testing kits to check for contaminants, ensuring the harvested rainwater is safe for its intended use.
5. Evaluate system efficiency by assessing the performance of various components and tracking the amount of water collected and used over time.
6. Troubleshoot common issues, such as decreased water pressure or leaks, and seek professional help when necessary to address more complex problems.
7. Regular maintenance can prevent many common issues and help ensure your rainwater harvesting system's longevity and optimal performance.
8. By following the tips and advice outlined in this chapter, you can contribute to a more sustainable and environmentally friendly way of life while enjoying the numerous benefits of a well-maintained and efficient rainwater harvesting system.

8

UTILIZING HARVESTED RAINWATER FOR VARIOUS HOUSEHOLD PURPOSES

A rain barrel overflowing with water, surrounded by lush green plants.

Rainwater harvesting is an ancient practice utilized by various civilizations throughout history. However, it is gaining **renewed attention** as a sustainable and environmentally friendly solution to address water scarcity and reduce our reliance on municipal water supplies. This chapter will explore the numerous ways harvested rainwater can be used within your home, making it an invaluable resource for daily living.

Rainwater is a versatile and abundant natural resource that can be collected, stored, and utilized for various household purposes. By implementing a rainwater harvesting system in your home, you can significantly reduce your water consumption, lower your utility bills, and contribute to conserving precious water resources. Moreover, using rainwater for various household tasks can help you maintain a healthier and more eco-friendly lifestyle.

The following sections will explore how to use harvested rainwater within your home, from irrigation and gardening to indoor cleaning and laundry applications. We will also discuss how to safely use rainwater for toilet flushing, bathroom use, cooking, and drinking, ensuring that you are well-equipped to maximize the benefits of your rainwater harvesting system.

As you read through this chapter, you will gain valuable insights into the practical applications of harvested rainwater and learn how to integrate it effectively into your daily routines. By the end, you will be inspired to harness the power of rainwater and transform your home into a more sustainable and eco-conscious haven. So, let's embark on this exciting journey and discover how harvested rainwater can enhance your household and improve your quality of life.

Irrigation and Gardening with Rainwater

Rainwater harvesting is an environmentally friendly practice and a practical solution for homeowners looking to reduce their water bills and maintain a lush, thriving garden. In this section, we will explore the numerous benefits of using harvested rainwater for irrigation and gardening purposes and provide tips on effectively utilizing this valuable resource.

One of the primary advantages of using harvested rainwater for irrigation is that it is **free of chemicals and minerals** commonly found in tap water. These substances can harm plants, as they may cause a buildup of salts in the soil, hindering plant growth and overall health. Rainwater, on the other hand, is naturally soft and free of these contaminants, making it an ideal choice for nourishing your garden.

In addition to being chemical-free, rainwater is an excellent source of nutrients for plants. As rainwater passes through the atmosphere, it collects nitrogen, a vital nutrient for plant growth. So when you use harvested rainwater for irrigation, you provide your plants with a natural fertilizer, promoting more robust and healthier growth.

To effectively utilize harvested rainwater for irrigation, it is essential to have a proper storage system in place. Rain barrels or cisterns can collect and store rainwater, which can then be distributed throughout your garden using a hose, drip irrigation system, or a simple watering can. When setting up your rainwater harvesting system, please place it in a location that allows easy access to your garden. Consider the size of your storage container based on your garden's water needs.

When using rainwater for irrigation, monitoring your garden's moisture levels is crucial to avoid overwatering. While rainwater is a valuable resource, providing your plants with too much water is still possible, which can lead to root rot and other issues. To maintain optimal moisture levels, consider investing in a soil moisture meter or use your finger to gauge the soil's dampness before watering.

In conclusion, utilizing harvested rainwater for irrigation and gardening is a sustainable and cost-effective way to maintain a healthy and vibrant garden. By providing your plants with chemical-free, nutrient-rich water, you promote their growth and contribute to a greener and more sustainable future. With the right storage system and proper watering techniques, you can maximize the benefits of rainwater harvesting and enjoy a flourishing garden all year round.

Indoor Cleaning and Laundry Applications

As we continue to explore the various ways harvested rainwater can be utilized within your home, let's delve into the world of indoor cleaning and laundry applications. By using rainwater for these tasks, you save on your water bill and contribute to a more sustainable and eco-friendly lifestyle. This section will discuss how rainwater can be effectively used for cleaning and laundry purposes, as well as the necessary precautions to ensure optimal results.

One of the most significant advantages of using rainwater for cleaning and laundry is its **softness.** Unlike hard water, which contains high levels of minerals like calcium and magnesium, rainwater is naturally soft. This means it is less likely to leave residue on surfaces, making it an excellent choice for cleaning tasks such as mopping floors, washing windows, and wiping down countertops. The absence of minerals in rainwater also means that it is gentle on your appliances, helping to prolong their lifespan and maintain their efficiency.

When it comes to laundry, using rainwater can have several benefits. Firstly, the softness of the water means that you can use less detergent, as it will lather more easily. This saves you money and reduces the chemicals released into the environment. Additionally, the lack of minerals in rainwater prevents the buildup of limescale in your washing machine, which can lead to reduced efficiency and costly repairs. Finally, clothes washed in rainwater tend to feel softer. They look brighter, as no minerals can dull the colors or leave a residue on the fabric.

To use harvested rainwater for indoor cleaning and laundry applications, it is essential to **ensure that the water is adequately filtered and debris-free.** While rainwater is generally clean, it can still pick up contaminants from the atmosphere or your roof and gutters. Therefore, it is crucial to have a proper filtration system in place to remove any particles and impurities. Additionally, using a separate pump and plumbing system for your rainwater supply is recommended to avoid cross-contamination with your main water supply.

In conclusion, utilizing harvested rainwater for indoor cleaning and laundry applications is an excellent way to save money, reduce your environmental impact, and enjoy the benefits of naturally soft water. By taking the necessary precautions and ensuring that your rainwater is adequately filtered, you can enjoy a cleaner, greener, and more efficient home.

Rainwater for Toilet Flushing and Bathroom Use

One of the most practical and efficient ways to utilize harvested rainwater in your home is for toilet flushing and bathroom use. Doing so can significantly reduce your household's potable water consumption and contribute to a more sustainable and eco-friendly lifestyle. In this section, we will explore the benefits of using rainwater for toilet flushing and bathroom use and provide guidance on how to set up a system for this purpose.

Toilet flushing accounts for a **substantial portion of a household's water usage**, with an average of 30% of indoor water consumption attributed to this activity. Therefore, harvesting rainwater can save potable water and reduce your water bill. Additionally, using rainwater for toilet flushing can help alleviate pressure on local water resources and reduce the energy required for water treatment and distribution.

Setting up a rainwater harvesting system for toilet flushing and bathroom use is relatively simple. First, you will need a storage tank to collect the rainwater. This tank should be connected to your home's gutter system, directing rainwater into the tank. Then, install a filter at the entrance of the tank to prevent debris and contaminants from entering the stored water.

Next, you will need a pump and a separate plumbing system to distribute the harvested rainwater to your toilets and bathroom fixtures. This system should be independent of your potable water supply to avoid cross-contamination. Ensuring the pump is powerful enough to provide adequate water pressure for your bathroom needs is also crucial.

To further enhance the water quality of the harvested rainwater, you can install a secondary filter or treatment system, such as a UV sterilizer or a chlorine dosing system. This step is crucial if you use rainwater for showering or bathing. In addition, it will help eliminate any potential health risks associated with using untreated rainwater.

In conclusion, using harvested rainwater for toilet flushing and bathroom use is an excellent way to conserve potable water and reduce your environmental footprint. By implementing a rainwater harvesting system in your home, you can enjoy the benefits of a more sustainable lifestyle while saving money on your water bill. In addition, with proper planning and installation, you can ensure that your rainwater harvesting system is safe, efficient, and effective in meeting your household's bathroom needs.

Cooking and Drinking: Ensuring Water Quality and Safety

As a homeowner who has embraced rainwater harvesting, you have already taken a significant step towards sustainable living and reducing your water footprint. While harvested rainwater can be used for various household purposes, it is essential to ensure its quality and safety, especially regarding cooking and drinking. In this section, we will discuss the necessary steps to guarantee that your harvested rainwater meets the required standards for consumption and culinary use.

First and foremost, it is crucial to maintain the cleanliness of your rainwater harvesting system. Regular inspection and cleaning of your catchment area, gutters, filters, and storage tanks will prevent the accumulation of debris, algae, and other contaminants that could compromise water quality. Additionally, it is advisable to install first-flush diverters to discard the initial flow of rainwater, which may contain higher levels of pollutants from the catchment surface.

Once you have ensured the cleanliness of your system, the next step is to treat the harvested rainwater to make it safe for consumption. Depending on your specific needs and preferences, several methods are available for treating rainwater. Some popular options include:

Filtration: Installing a multi-stage filtration system can effectively remove sediment, debris, and other suspended particles from the harvested rainwater. This process typically involves using sediment filters, activated carbon filters, and ultrafiltration or reverse osmosis membranes.

Disinfection: To eliminate harmful microorganisms such as bacteria, viruses, and protozoa, disinfection methods like chlorination, ultraviolet (UV) light treatment, or ozonation. These processes are highly effective in ensuring that your harvested rainwater is free from pathogens and safe for consumption.

Boiling: As a simple and cost-effective method, boiling your harvested rainwater for at least one minute can kill harmful microorganisms. However, boiling does not remove chemical contaminants or improve the taste and odor of the water.

Water testing: Regularly testing your harvested rainwater for contaminants is essential to ensure its safety for cooking and drinking. You can use home water testing kits or send samples to a certified laboratory for a comprehensive analysis. Based on the test results, you can adjust your treatment methods accordingly.

Once you have treated your harvested rainwater, it is ready for use in cooking and drinking. Using rainwater for these purposes reduces your reliance on municipal water supplies. In addition, it gives you a sense of satisfaction to know you are consuming a sustainable and eco-friendly resource. By following the guidelines discussed in this section, you can maximize the benefits of rainwater harvesting in your home while ensuring the health and well-being of your family.

Maximizing the Benefits of Rainwater Harvesting in Your Home

In conclusion, rainwater harvesting is an environmentally-friendly and cost-effective solution to address water scarcity and reduce the strain on our planet's limited resources. Implementing a rainwater harvesting system in your home can significantly decrease your dependence on municipal water supplies and contribute to a more sustainable future.

Throughout this chapter, we have explored various household applications for harvested rainwater, including irrigation and gardening, indoor cleaning and laundry, toilet flushing, bathroom use, cooking, and drinking. Each application offers unique benefits and opportunities for homeowners to maximize their harvested rainwater.

To maximize the benefits of rainwater harvesting in your home, it is essential to consider the following key points:

Choose the right system: Select a rainwater harvesting system that best suits your needs and budget. This may include factors such as the size of your property, the amount of rainfall in your region, and your specific water usage requirements.

Maintain your system: Regular maintenance of your rainwater harvesting system is crucial to ensure its efficiency and longevity. This includes cleaning gutters and filters, inspecting storage tanks, and monitoring water quality.

Optimize water usage: Be mindful of your water consumption habits and use harvested rainwater efficiently. This may involve using water-saving devices like drip irrigation systems and low-flow toilets and adopting water conservation practices in your daily routines.

Ensure water quality and safety: If you plan to use harvested rainwater for cooking and drinking, it is vital to implement appropriate treatment and filtration methods to guarantee the water's safety and quality. In addition, regular water supply testing and monitoring can help you maintain the highest water quality standards.

Educate and inspire others: Share your experiences and knowledge about rainwater harvesting with your friends, family, and community. By raising awareness about the benefits of rainwater harvesting, you can inspire others to adopt this sustainable practice and contribute to a greener future.

In summary, rainwater harvesting offers many benefits for homeowners, from reducing water bills to promoting self-sufficiency and environmental stewardship. By utilizing harvested rainwater for various household purposes and adopting best practices, you can maximize the potential of your rainwater harvesting system and enjoy the rewards of this sustainable solution for years to come.

Chapter Summary

1. Rainwater harvesting is an ancient practice that has gained renewed attention as a sustainable and environmentally friendly solution to address water scarcity and reduce reliance on municipal water supplies.
2. Harvested rainwater can be used for various household purposes, including irrigation and gardening, indoor cleaning and laundry, toilet flushing and bathroom use, and cooking and drinking, provided it is treated and filtered correctly.
3. Rainwater is naturally soft and free of chemicals and minerals, making it an ideal choice for healthy plants and maintaining a healthy garden.
4. Using rainwater for indoor cleaning and laundry applications can save money, reduce environmental impact, and prolong the lifespan of appliances.
5. Implementing a rainwater harvesting system for toilet flushing and bathroom use can significantly reduce household water consumption and contribute to a more sustainable lifestyle.
6. Ensuring the quality and safety of harvested rainwater for cooking and drinking is crucial. It can be achieved through proper system maintenance, filtration, disinfection, and regular water testing.
7. Maximizing the benefits of rainwater harvesting in your home involves choosing the right system, maintaining it, optimizing water usage, ensuring water quality and safety, and educating and inspiring others.
8. By adopting rainwater harvesting practices and utilizing harvested rainwater for various household purposes, homeowners can contribute to a greener future and enjoy the rewards of this sustainable solution.

9

LEGAL AND REGULATORY CONSIDERATIONS FOR RAINWATER HARVESTING

A rain barrel overflowing with water, surrounded by lush green plants.

Rainwater harvesting is environmentally friendly and cost-effective to conserve water, reduce your reliance on municipal water supplies, and lower utility bills. However, as more and more homeowners recognize the benefits of rainwater harvesting, they must be aware of the **legal and regulatory considerations** of implementing such systems at home. This chapter aims to provide you with a comprehensive understanding of the various rules and regulations governing rainwater harvesting, ensuring that your system is compliant and sustainable.

While rainwater harvesting has been practiced for centuries, modern-day systems are subject to various laws and regulations depending on your location. These rules are in place to protect public health, ensure your home's safety and structural integrity, and prevent potential conflicts with neighbors or homeowners associations. By familiarizing yourself with these legal and regulatory considerations, you can avoid fines, disputes, and other issues arising from non-compliance.

This chapter will explore the different aspects of legal and regulatory compliance for rainwater harvesting systems, including understanding local and state regulations, obtaining necessary permits and licenses, complying with building codes and standards, and navigating homeowners association rules and restrictions. By the end of this chapter, you will have a solid foundation of knowledge to help you ensure that your rainwater harvesting system is both legally compliant and environmentally sustainable.

Understanding Local and State Regulations

Rainwater harvesting is a sustainable and eco-friendly practice that has recently gained significant popularity. However, before embarking on your rainwater harvesting journey, it is crucial to understand the legal landscape that governs this practice. Local and state regulations significantly determine the feasibility and legality of rainwater harvesting for your home. This section will explore the importance of familiarizing yourself with these regulations and guide how to navigate them effectively.

First and foremost, it is essential to recognize that rainwater harvesting regulations can vary greatly depending on your location. Some states and municipalities have embraced this practice and actively encouraged it through incentives and lenient regulations. On the other hand, certain areas may have more stringent rules or even prohibit rainwater harvesting altogether. This variation in regulations underscores the importance of researching and understanding the specific laws that apply to your area.

To begin your research, consult your local government's website or contact the relevant department, such as the Department of Environmental Protection or Water Resources. These sources can provide valuable information on the legalities of rainwater harvesting in your area and any specific requirements or restrictions that may apply. It may also be helpful to consult with local experts or organizations specializing in rainwater harvesting, as they can offer firsthand knowledge and experience in navigating the regulatory landscape.

When researching local and state regulations, pay particular attention to the following aspects:

Legal status of rainwater harvesting: Determine whether rainwater harvesting is allowed in your area and if there are any restrictions on its use, such as for potable or non-potable purposes.

Rainwater collection methods: Some jurisdictions may have specific requirements or restrictions on the types of collection systems used, such as roof catchment systems or ground-level collection.

Water rights and ownership: In certain areas, water rights may be a complex issue, and rainwater harvesting could infringe on the rights of other water users. Be sure to understand the implications of water rights in your area and how they may affect your rainwater harvesting plans.

Water quality and treatment: Specific water quality standards and treatment requirements may be met depending on the harvested rainwater's intended use. Please familiarize yourself with these standards and ensure your planned system meets them.

Stormwater management and runoff: Some jurisdictions may have regulations to manage stormwater runoff and prevent potential flooding or environmental issues. Ensure your rainwater harvesting system complies with these regulations and does not negatively impact stormwater management efforts.

By taking the time to thoroughly understand the local and state regulations that govern rainwater harvesting, you can ensure that your system is both legally compliant and environmentally sustainable. In addition, this knowledge will help you avoid potential fines or penalties and contribute to the broader goal of promoting responsible water use and conservation in your community.

Obtaining Necessary Permits and Licenses

As you embark on your journey to integrate rainwater harvesting into your home, you must understand the permits and licenses required to ensure a smooth and legally compliant process. This section will guide you through the steps to obtain the necessary permits and licenses for your rainwater harvesting system.

First and foremost, it is crucial to research and familiarize yourself with the local and state regulations governing rainwater harvesting in your area. These regulations may vary significantly from location to location, so gathering **accurate and up-to-date information** is essential. You can consult your local government's website, contact your city or county's permitting office, or visit your local library to access resources.

Once you clearly understand the regulations, you can obtain the required permits and licenses. This may involve submitting an application to your local permitting office, providing detailed plans and specifications of your proposed rainwater harvesting system, and paying any associated fees. Be prepared to answer questions about the system's design, capacity, intended use, and any safety measures you plan to implement.

Sometimes, you may also need approval from other agencies or organizations, such as your local health department or water utility company. These organizations may have additional requirements or guidelines for rainwater harvesting systems, particularly if you plan to use the harvested water for potable purposes. Therefore, it is essential to communicate with these entities early in the process to avoid any delays or complications.

As you work through the permitting process, be prepared to adjust your rainwater harvesting system's design or specifications to meet regulatory requirements. This may involve modifying the system's capacity, incorporating additional filtration or treatment components, or implementing specific safety measures. It is essential to remain flexible and open to feedback from regulatory agencies to ensure your system is compliant and effective.

Finally, once you have obtained all necessary permits and licenses, it is crucial to keep them readily available and accessible. This will allow you to demonstrate compliance with local and state regulations should any questions or concerns arise. Also, maintaining accurate records of your permits and licenses can be helpful when selling your home or applying for insurance coverage, demonstrating your commitment to sustainable and responsible water management practices.

In conclusion, obtaining the necessary permits and licenses for your rainwater harvesting system is critical in ensuring compliance with local and state regulations. By thoroughly researching the requirements, engaging with regulatory agencies, and maintaining accurate records, you can confidently move forward with your rainwater harvesting project, knowing that you are contributing to a sustainable and environmentally friendly solution for your home.

Complying with Building Codes and Standards

This section will explore the importance of adhering to building codes and standards when implementing a rainwater harvesting system for your home. By understanding and complying with these regulations, you can ensure your rainwater harvesting system's safety, efficiency, and longevity.

Building codes and standards are established to provide a uniform set of guidelines for designing, constructing, and maintaining buildings and their various systems. These regulations are in place to ensure the safety and well-being of the occupants and protect the environment and surrounding community. Regarding rainwater harvesting, building codes and standards typically address the materials used, system design, installation, and maintenance.

To begin, it is essential to familiarize yourself with the specific building codes and standards for rainwater harvesting in your area. These regulations may vary depending on your location, so it is crucial to consult with your local building department or a knowledgeable professional to obtain accurate information. Additionally, it is essential to note that building codes and standards are subject to change, so staying up-to-date on any revisions is critical to maintaining compliance.

When designing your rainwater harvesting system, be sure to consider the following aspects that building codes and standards may regulate:

Materials: The materials used in your rainwater harvesting system must be durable, non-toxic, and suitable for their intended purpose. This may include using specific types of pipes, fittings, tanks, and filters that meet established standards.

System Design: The overall design of your rainwater harvesting system should adhere to guidelines that ensure its functionality and safety. This may involve requirements related to storage tanks' size and location, plumbing connections' configuration, and the incorporation of overflow and backflow prevention measures.

Installation: Proper installation is crucial to the performance and longevity of your rainwater harvesting system. Building codes and standards may dictate specific installation methods and require inspections and approval from local authorities.

Maintenance: Regular maintenance is necessary to keep your rainwater harvesting system operating efficiently and safely. Building codes and standards may outline recommended maintenance procedures and require periodic inspections and reporting to local authorities.

By complying with building codes and standards, you can ensure that your rainwater harvesting system is safe and efficient and contributes to the sustainable management of water resources in your community. The following section will discuss navigating homeowners association rules and restrictions that may impact your rainwater harvesting project.

Navigating Homeowners Association Rules and Restrictions

As you embark on your rainwater harvesting journey, you must be aware of and adhere to the rules and restrictions set forth by your local **Homeowners Association (HOA).** These organizations often have specific guidelines that govern the appearance and functionality of homes within their communities. In this section, we will discuss the importance of understanding your HOA's rules and restrictions and how to navigate them to ensure a successful and compliant rainwater harvesting system installation.

First and foremost, it is crucial to familiarize yourself with your HOA's governing documents, which typically include the Covenants, Conditions, and Restrictions (CC&Rs) and any additional rules and regulations specific to your community. These documents outline the standards and requirements for various aspects of your property, such as landscaping, exterior modifications, and sustainability initiatives like rainwater harvesting.

Consider any provisions impacting your rainwater harvesting plans when reviewing your HOA's rules and restrictions. For example, there may be guidelines regarding the size, location, and appearance of rain barrels or cisterns and requirements for screening or concealing these components from the public view. Additionally, your HOA may have specific rules about the materials that can be used for your rainwater harvesting system, such as requiring non-toxic, food-grade materials for water storage containers.

Suppose your HOA's rules and restrictions must be clarified or specifically address rainwater harvesting. In that case, contacting your HOA board or management company for clarification is a good idea. They can guide any existing policies or help you understand how your proposed system might fit within the community's overall aesthetic and sustainability goals. Sometimes, you may need to submit a formal request or application to your HOA to approve your rainwater harvesting system.

This process may require you to provide detailed information about your proposed system, including its design, materials, and intended use.

When working with your HOA, it is essential to maintain open lines of communication and demonstrate a willingness to collaborate. By being proactive and engaging in a constructive dialogue, you can educate your HOA about the benefits of rainwater harvesting and potentially pave the way for future installations within your community. Be prepared to address any concerns or questions. Consider sharing resources or examples of successful rainwater harvesting systems in other HOA-governed communities.

In conclusion, navigating your HOA's rules and restrictions is critical in implementing a rainwater harvesting system for your home. By understanding and adhering to these guidelines, you can ensure that your system is compliant and contributes to the overall sustainability and aesthetic appeal of your community. In addition, with careful planning and open communication, you can successfully integrate rainwater harvesting into your home while maintaining a positive relationship with your HOA.

Ensuring Compliance and Sustainable Rainwater Harvesting

In conclusion, the journey towards implementing a rainwater harvesting system for your home is not only about the technical aspects and environmental benefits but also about understanding and adhering to legal and regulatory considerations. By taking the time to familiarize yourself with local and state regulations, obtaining the necessary permits and licenses, complying with building codes and standards, and navigating homeowners association rules and restrictions, you can ensure that your rainwater harvesting system is both compliant and sustainable.

As a **responsible homeowner,** it is essential to recognize the importance of these legal and regulatory aspects, as they are in place to protect the environment, public health, and safety. Moreover, they help promote the responsible use of water resources and encourage adopting sustainable practices.

By following the guidelines and recommendations outlined in this chapter, you can confidently embark on your rainwater harvesting journey, knowing that you are taking the necessary steps to ensure compliance with the law and contribute to a more sustainable future for your community and the planet.

Remember, setting up a rainwater harvesting system for your home is not a one-time event but an ongoing commitment to sustainability and resource conservation. As you maintain and optimize your system, stay informed about any changes in regulations and best practices to ensure your efforts remain compliant and effective.

In the end, the rewards of rainwater harvesting go beyond the satisfaction of contributing to a greener environment and reducing your water bills. By taking the time to understand and adhere to legal and regulatory considerations, you also foster a sense of community and shared responsibility, inspiring others to follow in your footsteps and embrace sustainable living practices.

So, as you embark on this exciting and rewarding journey, remember that your efforts to harvest rainwater are beneficial for your home and wallet and play a crucial role in promoting a more sustainable and water-conscious future for generations to come.

Chapter Summary

1. Rainwater harvesting is an environmentally friendly and cost-effective way to conserve water. Still, knowing the legal and regulatory considerations of implementing such systems at home is essential.
2. Local and state regulations significantly determine the feasibility and legality of rainwater harvesting for your home. These regulations can vary greatly depending on your location.
3. Obtaining the necessary permits and licenses for your rainwater harvesting system is critical in ensuring compliance with local and state regulations.

4. Building codes and standards are established to provide a uniform set of guidelines for designing, constructing, and maintaining buildings and their various systems, including rainwater harvesting systems.
5. Homeowners Associations (HOAs) often have specific rules and restrictions that govern the appearance and functionality of homes within their communities, which may impact your rainwater harvesting plans.
6. Navigating your HOA's rules and restrictions is critical in implementing a rainwater harvesting system for your home. Maintaining open lines of communication with your HOA can help ensure a successful and compliant installation.
7. Ensuring compliance with legal and regulatory considerations is about protecting the environment, public health, and safety, promoting responsible water use, and adopting sustainable practices.
8. Setting up a rainwater harvesting system for your home is an ongoing commitment to sustainability and resource conservation. Staying informed about regulations and best practice changes is essential to ensure your efforts remain compliant and effective.

10

THE ENVIRONMENTAL AND ECONOMIC BENEFITS OF RAINWATER HARVESTING

A rain barrel overflowing with water, surrounded by lush green plants.

In today's rapidly changing world, where environmental concerns and resource management are on our minds, rainwater harvesting has emerged as a practical and sustainable solution for homeowners. This **age-old practice** of collecting and storing rainwater for later use has been gaining popularity in recent years as more people recognize its numerous environmental and economic benefits. By embracing rainwater harvesting, homeowners can contribute to a greener future and enjoy significant cost savings on their water bills.

Rainwater harvesting systems can range from simple rain barrels to more complex setups involving pumps, filters, and storage tanks. Regardless of the system's complexity, the core principle remains: capturing rainwater that would otherwise be lost as runoff and putting it to good use. This chapter will explore the various environmental and economic benefits of rainwater harvesting, including reducing the strain on municipal water supplies, minimizing the impact of stormwater runoff and soil erosion, promoting sustainable landscaping and gardening practices, saving money on water bills, and reducing energy consumption.

As you read through this chapter, you will gain a deeper understanding of how rainwater harvesting can positively impact both the environment and your wallet. Investing in a rainwater harvesting system for your home makes a wise financial decision. In addition, it takes a significant step towards a more sustainable and eco-friendly lifestyle. So, let's dive in and explore the myriad advantages of rainwater harvesting for your home and the environment.

Reducing the Strain on Municipal Water Supplies

As the world's population grows, the demand for clean, accessible water has increased. As a result, municipal water supplies are often stretched to their limits, struggling to meet the needs of both residential and commercial consumers. Homeowners can play a crucial role in alleviating this strain on our precious water resources by embracing rainwater harvesting.

Rainwater harvesting systems collect and store rainwater that would otherwise flow into storm drains, eventually returning to rivers, lakes, and oceans. By capturing this water, **homeowners can reduce their reliance on municipal water supplies**, freeing up resources for other essential services and uses. This is particularly important in areas that experience frequent droughts or water shortages, where every drop of water saved can make a significant difference.

The benefits of reducing the strain on municipal water supplies extend beyond the immediate water availability. As more people adopt rainwater harvesting practices, the need for costly infrastructure projects, such as dams and reservoirs, may be reduced. This can save taxpayers money and minimize the environmental impact associated with large-scale water management projects.

Furthermore, decreasing the demand for municipal water supplies can help protect our natural water sources from over-extraction. Many rivers, lakes, and aquifers are being depleted at an alarming rate, leading to devastating consequences for ecosystems and communities that rely on these water sources. By using harvested rainwater for non-potable purposes, such as irrigation and toilet flushing, homeowners can contribute to preserving these vital resources for future generations.

In summary, rainwater harvesting offers a practical and effective way for homeowners to reduce their dependence on municipal water supplies. By doing so, they can help conserve water resources, protect the environment, and support the long-term sustainability of our water supply systems.

Minimizing the Impact of Stormwater Runoff and Soil Erosion

One of the most significant yet often overlooked benefits of rainwater harvesting is its ability to **minimize the impact of stormwater runoff and soil erosion.** As urbanization expands, natural landscapes are replaced with impervious surfaces like concrete and asphalt. As a result, rain cannot be absorbed when it falls on these surfaces, leading to increased stormwater runoff. This excess water can cause many environmental issues, including flooding, soil erosion, and pollution of nearby water sources. We can play a vital role in mitigating these problems by implementing rainwater harvesting systems in our homes.

Stormwater runoff is a significant concern in urban areas, as it can overload storm drains and sewer systems. This can result in localized flooding, causing damage to property and infrastructure. By capturing rainwater before it becomes runoff, we can significantly reduce the volume of water entering these systems. This helps prevent flooding and decreases the number of pollutants washed into our rivers, lakes, and oceans.

In addition to reducing the strain on stormwater management systems, rainwater harvesting can also help to minimize soil erosion. When rain falls on natural landscapes, it is absorbed by the soil and vegetation, which helps maintain the land's stability. However, when rain falls on impervious surfaces, it can wash away the topsoil, leading to erosion. This can have devastating consequences for both the environment and the economy, as it can lead to the loss of fertile land, damage to infrastructure, and increased sedimentation in waterways.

By capturing rainwater and using it to irrigate our gardens and landscapes, we can help maintain the soil's health and stability. This not only prevents erosion but also promotes the growth of vegetation, which can further help to absorb and filter rainwater. Moreover, using harvested rainwater for irrigation can reduce the demand for potable water, often sourced from rivers and lakes. This can help to maintain healthy ecosystems and protect the habitats of various plant and animal species.

In conclusion, rainwater harvesting offers a practical and effective solution for minimizing the impact of stormwater runoff and soil erosion. By capturing and utilizing rainwater, we can reduce the strain on our stormwater management systems, protect our water sources from pollution, and maintain the health and stability of our landscapes. As we continue to embrace the advantages of rainwater harvesting, we can look forward to a more sustainable and resilient future for our homes and the environment.

Promoting Sustainable Landscaping and Gardening Practices

As we continue to explore the environmental and economic benefits of rainwater harvesting, it's essential to highlight its role in promoting sustainable landscaping and gardening practices. With the increasing awareness of the importance of eco-friendly approaches, rainwater harvesting is vital for homeowners looking to create a greener environment around their homes.

Sustainable landscaping refers to designing, constructing, and maintaining outdoor spaces that minimize resource consumption, reduce waste, and preserve the natural environment. Gardening practices, on the other hand, focus on nurturing plants and vegetation in an eco-friendly manner. In addition, homeowners can significantly contribute to a healthier and more sustainable ecosystem by incorporating rainwater harvesting into these practices.

One of the primary ways rainwater harvesting supports sustainable landscaping is by providing a **natural and chemical-free water source for plants.** Unlike treated municipal water, rainwater is free from chemicals such as chlorine and fluoride, which can harm plants and soil health. This means that by using harvested rainwater, you are conserving water resources and promoting a healthier environment for your plants to thrive.

Moreover, rainwater harvesting enables homeowners to maintain their gardens and landscapes during water restrictions or droughts. This is particularly important in regions where water scarcity is a pressing issue. By having a rainwater harvesting system, you can continue caring for your plants without relying on limited municipal water supplies.

Another advantage of using harvested rainwater for landscaping and gardening is its positive impact on soil quality. Rainwater is naturally soft, meaning it has a lower mineral content than hard water. This characteristic of rainwater helps prevent the buildup of salts in the soil, which can be detrimental to plant growth and soil structure. As a result, using rainwater for irrigation promotes a healthier soil environment that encourages plant growth and reduces the need for chemical fertilizers.

Lastly, sustainable landscaping and gardening practices often involve the use of native plants that are well-adapted to local climate conditions. These plants typically require less water and maintenance than non-native species, making them an ideal choice for eco-conscious homeowners. By combining native plants with rainwater harvesting, you can create a low-maintenance and water-efficient landscape that is both beautiful and environmentally friendly.

In conclusion, rainwater harvesting plays a significant role in promoting sustainable landscaping and gardening practices. By providing a natural, chemical-free water source, reducing the strain on municipal water supplies, and supporting soil health, rainwater harvesting allows homeowners to create a greener and more sustainable environment around their homes. Embracing these practices benefits the environment and contributes to a healthier and more vibrant outdoor space for you and your family to enjoy.

Saving Money on Water Bills and Reducing Energy Consumption

One of the most tangible benefits of rainwater harvesting is the potential to **save money on your water bills.** As water becomes an increasingly scarce resource, its cost rises. Collecting and utilizing rainwater can significantly reduce your reliance on municipal water supplies and decrease your monthly expenses.

Rainwater is a free resource that falls from the sky, making it an attractive alternative to purchasing water from your local utility company. Installing a rainwater harvesting system allows you to capture this valuable resource for various household purposes, such as watering your garden, flushing toilets, and even washing clothes. In many cases, rainwater can be used as a direct substitute for tap water, allowing you to save money without sacrificing convenience or quality.

In addition to direct financial savings, rainwater harvesting can also help you reduce your energy consumption. Treating and distributing water requires a significant amount of energy, contributing to greenhouse gas emissions and climate change. Using rainwater instead of relying solely on municipal water supplies can help reduce the energy required to pump and treat water, thereby decreasing your carbon footprint.

Furthermore, using rainwater can help reduce the demand for energy-intensive water heating. Rainwater is typically cooler than tap water, requiring less energy to heat up for various household uses. This can result in lower energy bills and a reduced environmental impact.

In conclusion, rainwater harvesting offers a multitude of financial and environmental benefits. By capturing and utilizing this free and abundant resource, you can save money on your water bills, reduce energy consumption, and contribute to a more sustainable future.

Investing in a Greener Future with Rainwater Harvesting

In conclusion, investing in rainwater harvesting systems for your home is a **wise financial decision** and a significant step towards a greener and more sustainable future. As we have explored throughout this chapter, the environmental and economic benefits of rainwater harvesting are numerous and far-reaching.

By reducing the strain on municipal water supplies, homeowners who embrace rainwater harvesting contribute to conserving precious water resources and help alleviate the pressure on water treatment facilities. This, in turn, leads to a decrease in energy consumption and a reduction in greenhouse gas emissions.

Furthermore, rainwater harvesting systems are crucial in minimizing the impact of stormwater runoff and soil erosion. By capturing and storing rainwater, we can prevent the harmful effects of uncontrolled water flow on our ecosystems, including the destruction of habitats and the pollution of waterways.

In addition to these environmental benefits, rainwater harvesting promotes sustainable landscaping and gardening practices. Using harvested rainwater for irrigation allows homeowners to maintain lush and healthy gardens without relying on chemically treated municipal water. This benefits the plants and supports the overall health of the surrounding environment.

Lastly, the economic benefits of rainwater harvesting must be balanced. By reducing reliance on municipal water supplies, homeowners can save considerable money on their water bills. Moreover, the reduced energy consumption associated with water treatment and distribution translates to lower energy costs and a decreased carbon footprint.

In summary, rainwater harvesting is an investment in a greener future for both individual homeowners and the planet as a whole. By adopting this sustainable practice, we can collectively work towards a more environmentally responsible and economically viable future, ensuring future generations' health and well-being. So, take the first step towards a greener future and start reaping the rewards of rainwater harvesting today.

Chapter Summary

1. Rainwater harvesting is a practical and sustainable solution for homeowners, offering numerous environmental and economic benefits, such as reducing strain on municipal water supplies, minimizing stormwater runoff and soil erosion, promoting sustainable landscaping and gardening practices, and saving money on water bills and energy consumption.
2. By capturing and storing rainwater, homeowners can reduce their reliance on municipal water supplies, freeing up resources for other essential services and uses and potentially reducing the need for costly infrastructure projects like dams and reservoirs.
3. Rainwater harvesting helps minimize the impact of stormwater runoff and soil erosion by capturing rainwater before it becomes runoff, reducing the volume of water entering stormwater management systems, and preventing the washing away of topsoil.
4. Sustainable landscaping and gardening practices can be supported by rainwater harvesting, as it provides a natural, chemical-free source of water for plants, maintains soil health, and allows for native plants that require less water and maintenance.

5. Rainwater harvesting can save homeowners money on their water bills by reducing their reliance on municipal water supplies and utilizing rainwater for various household purposes, such as watering gardens, flushing toilets, and washing clothes.
6. Energy consumption can be reduced through rainwater harvesting, as it decreases the energy required to pump and treat water. Using rainwater can also reduce the demand for energy-intensive water heating.
7. Rainwater harvesting contributes to a greener and more sustainable future by conserving water resources, reducing energy consumption, and promoting eco-friendly landscaping and gardening practices.
8. Investing in rainwater harvesting systems is a wise financial decision and a significant step towards a more environmentally responsible and economically viable future for both individual homeowners and the planet as a whole.

EPILOGUE

As we conclude this enlightening exploration into the world of rainwater harvesting, we must pause and reflect on our journey together. From understanding the basic principles of rainwater harvesting to delving into the intricacies of various systems, we have uncovered the immense potential of this simple yet powerful practice. By embracing rainwater harvesting for our homes, we are taking a step towards self-sufficiency and contributing to a more sustainable and environmentally conscious future.

Throughout this book, we have seen how **rainwater harvesting can profoundly impact our environment and communities.** By reducing the strain on our water resources and decreasing the demand for energy-intensive water treatment processes, we are actively participating in preserving our planet. Moreover, we have learned that rainwater harvesting can foster community and shared responsibility as neighbors come together to implement and maintain these systems.

In our exploration, we have also addressed the challenges and misconceptions surrounding rainwater harvesting. By dispelling myths and providing accurate information, we have empowered ourselves with the knowledge to make informed decisions and confidently advocate for the adoption of rainwater harvesting in our communities.

We have also discussed the crucial role that government, industry, and individuals must play in promoting rainwater harvesting. Through policy changes, incentives, and education, we can create an environment where rainwater harvesting becomes the norm rather than the exception. By working together, we can ensure that the benefits of rainwater harvesting are accessible to all, regardless of socioeconomic status or geographical location.

As we envision a world where rainwater harvesting is commonplace, we can begin to imagine the countless benefits that would follow. From healthier ecosystems to stronger communities, the widespread adoption of rainwater harvesting would undoubtedly contribute to a more sustainable and resilient future.

In the following sections, we will delve deeper into the impact of rainwater harvesting on our environment and communities, explore the challenges that must be overcome, and discuss the roles that various stakeholders must play in promoting this practice. Finally, we will share our thoughts on the path forward for rainwater harvesting and how it can contribute to a sustainable future for all.

The Impact of Rainwater Harvesting on Our Environment and Communities

As we reach the end of our journey exploring rainwater harvesting, it is essential to take a moment to reflect on the profound impact this practice can have on our environment and communities. Rainwater harvesting is not just a means to conserve water but a **powerful tool** that can contribute to a more sustainable and resilient future for all.

First and foremost, rainwater harvesting helps to alleviate the stress on our overburdened water resources. With increasing population growth, urbanization, and climate change, the demand for freshwater is rapidly outstripping the supply. By capturing and utilizing rainwater, we can reduce reliance on traditional water sources, such as rivers, lakes, and groundwater, which are already under immense pressure. This, in turn, helps to preserve these vital resources for future generations and protect the delicate ecosystems that depend on them.

Moreover, rainwater harvesting can play a significant role in mitigating the effects of climate change. As extreme weather events, such as floods and droughts, become more frequent and severe, the need for adaptive water management strategies becomes increasingly urgent. Rainwater harvesting systems can help buffer against these impacts by providing an alternative water source during times of scarcity and reducing the risk of flooding by slowing down and storing stormwater runoff.

In addition to its environmental benefits, rainwater harvesting can have a transformative effect on our communities. By promoting self-sufficiency and resilience, rainwater harvesting empowers individuals and neighborhoods to take control of their water supply and reduce their dependence on centralized water systems. This fosters a sense of pride and ownership and encourages a more responsible and mindful approach to water use.

Furthermore, rainwater harvesting can help to bridge the gap between water-rich and water-poor communities. In many parts of the world, access to clean, safe water remains a luxury rather than a fundamental human right. By implementing rainwater harvesting systems, these communities can tap into a previously underutilized resource and improve their water security, health, and overall quality of life.

Lastly, the widespread adoption of rainwater harvesting can stimulate local economies by creating new job opportunities in designing, installing, and maintaining these systems. This, in turn, can contribute to the growth of a green economy, which prioritizes environmental sustainability and social equity.

In conclusion, the impact of rainwater harvesting on our environment and communities is profound and far-reaching. Yet, by embracing this practice, we can work together to create a more sustainable, resilient, and equitable future for all.

Overcoming Challenges and Misconceptions Surrounding Rainwater Harvesting

As we have explored the world of rainwater harvesting throughout this book, addressing the challenges and misconceptions that often arise when discussing this sustainable practice is essential. By overcoming these hurdles, we can pave the way for a future where rainwater harvesting is widely adopted and appreciated for its numerous benefits.

One common misconception is that rainwater harvesting is a complicated and expensive process. While it is true that some systems can be quite ***intricate***, there are also many simple and affordable options available for homeowners. From basic rain barrels to more advanced systems, there is a rainwater harvesting solution for every budget and level of expertise. Educating ourselves and others about the various options can dispel the myth that rainwater harvesting is only for the wealthy or highly skilled.

Another challenge faced by proponents of rainwater harvesting is the concern about water quality. Some individuals worry that harvested rainwater may be unsafe for use, particularly for drinking. However, with proper filtration and treatment, rainwater can be as safe as other sources. Therefore, it is crucial to emphasize the importance of regular maintenance and the use of appropriate filters to ensure the safety and quality of harvested rainwater.

A further misconception is that rainwater harvesting is only suitable for regions with abundant rainfall. In reality, rainwater harvesting can be a valuable resource even in areas with limited precipitation. By implementing efficient collection and storage methods, homeowners in drier climates can still reap the benefits of rainwater harvesting. Moreover, utilizing harvested rainwater can help alleviate the strain on water resources in regions where water scarcity is a pressing issue.

To overcome these challenges and misconceptions, raising awareness and promoting education about rainwater harvesting is vital. Government agencies, industry leaders, and individuals all have a role in spreading accurate information and encouraging adoption of this sustainable practice. In addition, by offering incentives, sharing success stories, and providing resources for learning, we can foster a culture that embraces rainwater harvesting as a viable and beneficial solution for our water needs.

In conclusion, overcoming the challenges and misconceptions surrounding rainwater harvesting is crucial in promoting its widespread adoption. By dispelling myths, addressing concerns, and providing accurate information, we can pave the way for a future where rainwater harvesting is embraced as a sustainable and valuable practice for homes and communities worldwide.

The Role of Government, Industry, and Individuals in Promoting Rainwater Harvesting

As we have explored throughout this book, rainwater harvesting offers numerous benefits for our environment, communities, and households. However, to truly maximize its potential and make it a widespread practice, the collective efforts of government, industry, and individuals are crucial. Therefore, this section will delve into each stakeholder's role in promoting rainwater harvesting and fostering a sustainable future.

First and foremost, the **government** plays a pivotal role in creating a supportive framework for rainwater harvesting. By implementing policies and regulations encouraging its adoption, the government can create a favorable environment for individuals and businesses to invest in rainwater harvesting systems. This can be achieved through various means, such as offering financial incentives, tax breaks, or rebates for those who install rainwater harvesting systems. Additionally, incorporating rainwater harvesting into building codes and urban planning guidelines can promote integration into new construction projects and community developments.

On the other hand, the **industry** is responsible for innovating and improving rainwater harvesting technologies, making them more efficient, affordable, and accessible to a broader audience. Manufacturers and suppliers should develop high-quality, user-friendly systems catering to different needs and budgets. Furthermore, industry stakeholders should collaborate with educational institutions and research organizations to advance the knowledge and understanding of rainwater harvesting and train the next generation of professionals in this field.

Lastly, **individuals** play a vital role in promoting rainwater harvesting by embracing it as a sustainable practice in their daily lives. Homeowners and property owners can invest in rainwater harvesting systems, reaping the benefits of reduced water bills and a more eco-friendly lifestyle. Moreover, individuals can advocate for rainwater harvesting by sharing their experiences and knowledge with their communities, friends, and family members. This grassroots approach can help to dispel misconceptions and raise awareness about the advantages of rainwater harvesting.

In conclusion, the successful promotion of rainwater harvesting relies on the combined efforts of government, industry, and individuals. By working together, these stakeholders can create a world where rainwater harvesting is the norm, contributing to a more sustainable and water-secure future for all.

Envisioning a World Where Rainwater Harvesting is the Norm

Imagine a world where every home, office building, and public space has an efficient rainwater harvesting system. A world where the sight of rainwater tanks, catchment systems, and green roofs is as commonplace as solar panels and wind turbines. This is the world we strive for, where rainwater harvesting is not just an alternative but the norm.

In this world, we would witness a **dramatic reduction in the demand for treated water from municipal sources**, leading to decreased energy consumption and a reduced carbon footprint. In addition, our cities would be transformed into sustainable urban landscapes where green spaces and rainwater harvesting systems work harmoniously to reduce the urban heat island effect and improve air quality.

The widespread adoption of rainwater harvesting would also lead to a more resilient water supply, especially in regions prone to drought or water scarcity. In addition, by relying on a decentralized water source, communities would be better equipped to handle periods of low rainfall and reduced water availability. This would not only benefit the environment but also help alleviate the economic burden on households and businesses that often face skyrocketing water bills during water scarcity.

Moreover, embracing rainwater harvesting as the norm would foster a greater sense of environmental stewardship among individuals. As people become more aware of their water consumption and its impact on the planet, they will be more inclined to adopt other sustainable practices daily. This shift in mindset could lead to a ripple effect, inspiring further innovation and collaboration in pursuing a more sustainable future.

To achieve this vision, it is essential for rainwater harvesting to be integrated into the design and construction of new buildings and infrastructure projects. Architects, engineers, and urban planners must work together to develop innovative solutions that seamlessly incorporate rainwater harvesting systems into the fabric of our built environment. Additionally, retrofitting existing structures with rainwater harvesting technology should be encouraged and incentivized, making it accessible to more property owners.

In this envisioned world, rainwater harvesting would no longer be seen as a niche or alternative practice but as an integral component of a sustainable lifestyle. By embracing rainwater harvesting as the norm, we can collectively work towards a future where our water resources are managed responsibly, our environment is protected, and our communities are more resilient in the face of a changing climate.

Final Thoughts: The Path Forward for Rainwater Harvesting and a Sustainable Future

As we reach the end of our exploration into the world of rainwater harvesting, it is essential to take a moment to reflect on the significance of this practice in our pursuit of a sustainable future. Throughout this book, we have delved into the **numerous benefits**, **challenges**, and **opportunities** that rainwater harvesting presents. This practice could revolutionize how we manage our water resources.

The path forward for rainwater harvesting is collaboration, innovation, and determination. By working together, governments, industries, and individuals can create a world where rainwater harvesting is accepted and celebrated as a vital component of sustainable living. This will require continued efforts to educate the public about the benefits of rainwater harvesting, dispel misconceptions, and encourage adopting this practice in homes and communities worldwide.

Innovation will also play a crucial role in the future of rainwater harvesting. As technology advances, we can expect new and improved systems that make rainwater harvesting more efficient and accessible. This will make it easier for individuals to incorporate rainwater harvesting into their daily lives and help drive down costs and increase the overall appeal of this practice.

As individuals, we must also recognize our role in promoting rainwater harvesting and sustainable living. By embracing this practice in our homes and sharing our experiences with others, we can create a ripple effect that will ultimately lead to the widespread adoption of rainwater harvesting. This, in turn, will have a profound impact on our environment, our communities, and our planet as a whole.

In conclusion, the journey of rainwater harvesting is far from over. As we look to the future, we must continue to champion this practice and work together to create a world where rainwater harvesting is a common practice and a cornerstone of sustainable living. By doing so, we can help to ensure that future generations inherit a planet that is healthy, vibrant, and capable of sustaining life for many years to come. So, let us embrace the potential of rainwater harvesting and take the first steps toward a more sustainable future together.

ACKNOWLEDGMENTS

I want to express my sincere gratitude to all of the individuals who have supported me throughout the writing of this book.

First and foremost, I thank my family for their unwavering support and encouragement. Their love and belief in me have kept me going during the many long nights spent researching and writing.

I also want to thank my colleagues and mentors, who have generously shared their knowledge and expertise with me. Their guidance and support have been invaluable in helping me to understand and present the complex concepts contained in this book.

Finally, I thank my editor and publisher for their patience and hard work in bringing this book to fruition. Their dedication to producing a high-quality product is greatly appreciated.

To all of these people and more, thank you for helping to make this book a reality.

ABOUT THE AUTHOR

Michael Barton is an expert in sustainability and regenerative agriculture with over a decade of experience. He holds a degree in Environmental Studies and has received advanced training in the principles and practices of regenerative agriculture.

Throughout his career, Michael has worked with farmers, policymakers, and environmentalists to promote sustainable farming practices and advocate for the adoption of regenerative agriculture. He has collaborated with organizations worldwide to advance sustainable agriculture and food systems.

In his free time, Michael enjoys hiking and spending time with his family.

Made in the USA
Monee, IL
14 November 2025

34889450R00079